PRENTICE-HALL FOUNDATIONS
OF MODERN BIOLOGY SERIES

WILLIAM D. MCELROY
AND CARL P. SWANSON, *editors*

ANIMAL BEHAVIOR

3rd edition

V. G. DETHIER
Professor of Biology, Princeton University

ELIOT STELLAR
Professor of Physiological Psychology, University of Pennsylvania

PRENTICE-HALL, INC.

ENGLEWOOD CLIFFS, NEW JERSEY

FOUNDATIONS OF MODERN BIOLOGY SERIES   WILLIAM D. MCELROY
AND CARL P. SWANSON, *editors*

C—13–037465–2
P—13–037440–7
Library of Congress Catalog Card Number   78–110092

Current printing   10 9 8 7 6 5 4 3 2 1

PRENTICE-HALL INTERNATIONAL, INC.,
PRENTICE-HALL OF AUSTRALIA, PTY. LTD.,
PRENTICE-HALL OF CANADA, LTD.,
PRENTICE-HALL OF INDIA PRIVATE LTD.,
PRENTICE-HALL OF JAPAN, INC.,

THIS SERIES, FOUNDATIONS OF MODERN BIOLOGY, WHEN LAUNCHED A number of years ago, represented a significant departure in the organization of instructional materials in biology. The success of the series provides ample support for the belief, shared by its authors, editors, and publisher, that student needs for up-to-date, properly illustrated texts and teacher prerogatives in structuring a course can best be served by a group of small volumes so planned as to encompass those areas of study central to an understanding of the content, state, and direction of modern biology. The twelve volumes of the series still represent, in our view, a meaningful division of subject matter.

This edition thus continues to reflect the rapidly changing face of biology; and many of the consequent alterations have been suggested by the student and teacher users of the texts. To all who have shown interest and aided us we express thankful appreciation.

WILLIAM D. MCELROY
CARL P. SWANSON

BOOKS ON ANIMAL BEHAVIOR CAN BE WRITTEN FROM MANY DIFFERENT points of view, and indeed they have been. Some emphasize investigations of the vertebrates most often used by laboratory workers in psychology and physiology: the rat, the cat and dog, and the monkey. Some report the rich observations of naturalists and biological field workers. Some are concerned with social behavior, some with learning. All contriblute to the broad understanding of animal behavior and all, of course, omit much.

This book has its own emphasis and its own omissions, but it represents an attempt on the part of the authors to develop meaningful and interesting ways of looking at animal behavior. In the first place, our viewpoint is evolutionary, for we see much to be gained by looking at behavior as an adaptive mechanism over the whole phylogenetic scale, from ameba to man. Secondly, our viewpoint is neurological, for we believe that behavior is best understood as a function of the activity of the nervous system.

As the nervous system changes in phylogeny, so does the behavior. The modes of adaptation change from the innate, stereotyped patterns to the highly flexible and variable acquired patterns. Thus, we believe the key to understanding behavior lies in our understanding of the evolution of the nervous system.

The evolutionary and neurological approach to animal behavior, of course, includes the study of man, the most recently evolved animal, with the most intricate nervous system and the most complex behavior. The hope is that the principles we learn in the study of the simpler organisms will help us understand the emotion and the intelligence of man and the successes and failures of his societies.

V. G. DETHIER
ELIOT STELLAR

CONTENTS

NO ANIMAL LIVES ALONE. EACH COMES INTO CONTACT WITH OTHER animals sometime in its life. That most solitary of creatures, the albatross, which spends the greater part of its life at sea far out of sight of land, returns periodically to the company of its own to breed. Even animals whose reproductive needs can be satisfied in solitude—the ameba that divides by fission, the budding *Hydra*, the hermaphroditic snail, the parthenogenetic aphid—are not alone. They have contact with the organisms that constitute their food and the parasites and predators that prey upon them.

Thus, few animals are solitary. Many are drawn together in fluid temporary associations by shifting conditions of the environment, such as the lee of a stone in a running brook, a rotting log in the jungle, a water hole in the desert, a carcass in the veld. Some are gathered in more unified communities, as shoals of fishes and flocks of birds; some live in highly integrated colonies, as do ants and honeybees; others (such as the corals) are so intimately associated that it is impossible to tell the individual from the colony.

Regardless of the closeness with which animals associate, their relations are vastly different from aggregations of nonliving entities such as molecules. Although the movements of a flock of starlings or a crowd of ants on a broken anthill can perhaps be analyzed in the same mathematical terms as are applied to Brownian molecular movement, the two are fundamentally different, and the difference is not merely a matter of complexity. The actions of animals are directed toward keeping the animals alive and enabling them to reproduce. Thus, the continued existence of an organism and of a species depends on the effectiveness of actions of the individual. This is not true of molecules. The directive action of organisms with respect to one another is one aspect of behavior.

Survival also depends on the maintenance of suitable relations with the nonliving environment. The sparrow fluffs its feathers when it is cold, the locust orients its body toward the sun when it is hot, the bat hibernates during winter, the swallow flies south. This adaptive relation between an organism and its environment is also *behavior*.

The assertion that an organism maintains relations with its environment, whether nonliving or living, implies that an organism changes in response to changes in the environment. These changes, which we call behavior, are not passive; they are directed actions, that is, actions promoting survival, and they are reversible. When an oak tree bends and thrashes in the wind, it is a passive thing. When a parasitic plant such as the dodder reaches out and twines itself around another plant, it does so by growth movements that are irreversible. By contrast, the behavior of animals is both active and reversible.

The Venus-flytrap (*Dionaea muscipula*) is an exception to this generalization. On the surface of the open trap there are mechanosensory hairs that trigger the closing response. When an alighting insect touches these hairs, a receptor potential is generated, which in turn initiates an action potential (see page 11). The trap now closes, the whole action having required about 80 msec (milliseconds). One or two weeks later the trap reopens. If the prey has escaped capture, the trap reopens within 12 to 44 hr. Electrophysiologically as well as behaviorally, this plant resembles a simple animal.

The capacity to respond to stimuli is termed *irritability* and is an essential property of living matter. Responses that do not jeopardize survival or that promote survival of the young even though jeopardizing survival of the parent are preserved in the course of time; others are lost. In the long developmental history of animals, as cells became associated into tissues, tissues into organs, and organs into organisms, the suitability of the changes that took place in each individual cell, tissue, or organ began to be measured, not solely in terms of the individual unit, but in terms of the survival requirements of the whole. The cell that responded unsuitably was eliminated, as was also the deficient tissue or organ. In short, the course of the evolution of responsiveness has been shaped by the requirements of the whole individual, acting in his environment. Since no animal is ever able to free itself completely from its heredity, its behavior must be seen as inevitably coupled to its evolutionary history.

The change from a unicellular plan of construction to a multicellular one permitted animals to become larger. With multicellularity came a greater need for coordination; with the increase in size came a greater need for the conduction of information within the organism. To meet this need, certain cells of the multicellular organism specialized in enhancing both its irritability and the rapid conduction of changes in irritability through the animal. With further specialization, these cells became so organized that there grew up a division of labor to detect environmental change (both inside and outside the organism), to conduct information, to integrate information, and to initiate a response. These cells became the nervous system, the matrix of behavior. It follows that an organism's behavior is an expression principally of the capabilities of its nervous system.

The study of animal behavior is, therefore, an analysis of the potentialities of the nervous system. Note that we say "potentialities" rather than "physiology" of the nervous system since we want to emphasize that the nervous system affects the whole animal or an organ system of the animal and controls the animal's changing relations with its environment. The nervous system does not operate in a vacuum. It is affected by the limitations and biases imprinted upon it during its long evolutionary history, by its particular stage of development in the growing individual, by the changes impressed upon it by its own repeated performance, and by the influence of internal and external environmental changes. Ultimately, its actions are translated into some effect outside itself.

The main function of the nervous system is to control the release of energy through the contraction of muscles and the secretions of glands; it also regulates such other forms of energy as light (the firefly) and electromotive force (the electric eel). Of these effects, muscular movement is surely the most widespread and important. Until complex muscular systems evolved, intricate behavior was impossible even though the nervous system may have possessed the potentiality for such behavior. The control of muscular energy involves more than mere efficiency of contractile mechanisms. The contractions of the various muscles must be synchronized so that their work is directed to serve the requirements of the animal. Otherwise there is a useless waste of energy. Muscular movements differ widely in complexity: in frequency, in sequential relationships, in duration, in intensities, in overall patterns of action, in their influence on other movements, in the circumstances that initiate them, and in reactions to previous effects.

To understand behavior we must look first at certain aspects of the nervous system. We must examine its role in recording changes in the environment, in assessing this information, and in coordinating the various muscular activities that will best serve the economical operation of the animal. But our search for understanding will not end with the nervous system, because we shall discover that the nervous system, in activating the muscles, the glands, and other effectors, indirectly acts upon itself—for example, when part of the nervous system causes a muscle to contract, the action of the muscle stimulates other nerves, which then may terminate the initial action. Our search, therefore, will extend to these other systems, the glands and muscles, for it is from the interplay of all these systems that behavior is made.

FROM PROTOPLASMIC IRRITABILITY TO COGNITION IS A DEVELOPMENT
that has required upwards of a billion years. This development is in-
extricably bound to the evolution of the nervous system. In the beginning, of course, there was no nervous system. Isolated masses of living
material probably showed general and uniform irritability. Whenever
a stimulus caused a change at one point, this change undoubtedly
stimulated adjacent areas so that a wave of excitation spread slowly in
all directions, like the ripples from a stone dropped in a quiet pool.

Through the course of time, the property of irritability gradually
tended to be localized, channeled, and refined so that the organism was
no longer equally excitable throughout. The conduction of excitation
was speeded and channeled into particular pathways. The advantages of
this refinement are clear, since otherwise the organism would be a constantly reacting thing lacking versatility and finesse of discrimination.
In the course of evolution, irritability has been channeled in two broad
directions: In one, elaboration took place at a subcellular level within
the confines of a single plasma membrane; in the other, elaboration
occurred at a cellular level, and a multicellular structure developed.
Throughout the animal kingdom there are many examples of different
structural levels of nervous organization.

In the protozoa of today, great differences in complexity exist, from the relatively undifferentiated protoplasmic mass of an ameba to the elaborate organelle system of the ciliates, for example, *Paramecium*. The protoplasm in the ciliates consists of areas specialized for detecting changes in the environment, other areas designed for conducting excitation to various parts of the body, and still others for producing limited and local responses. These specialized areas include sensory bristles, photoreceptors, cilia for swimming, cirri (fused cilia) for crawling, food-catching devices, organelles for attachment, trichocysts (microscopic harpoonlike structures believed to assist in the holding of prey), contractile fibrils (*myonemes*) that may even be cross-striated, and fibrils whose function is presumed to be conduction. Figure 1.1 illustrates an experiment designed to show the function of these fibrils.

We do not fully understand the actions of these relatively elaborate systems, but in them irritability is essentially part of a single structure. The inherent intimacy of the structure sets limits on the complexity of the mechanism, on the scope of its performance, and, above all, on its potentiality.

Acellular systems are, however, obviously capable of serving the primary needs of the animal for food, protection, and reproduction. To fulfill these needs, the animal must be able to move and to detect environmental features. All these functions may be performed by relatively undifferentiated protoplasm, as is demonstrated by an ameba. The ameba's sensory world seems to be divided into food and nonfood. Virtually all stimuli that are not food elicit withdrawal reactions; food

Figure 1.1 *When an incision (a) is made in the ciliate Euplotes in such a way as to cut the fibrils (f) running from the motorium (m)—spot where activity originates—to the cirri (c), coordination of locomotion is lost. Other incisions (b) do not interfere with locomotion.* [Redrawn from C. V. Taylor, *Univ. Calif. Publ. Zool., 19* (1920), 403–470.]

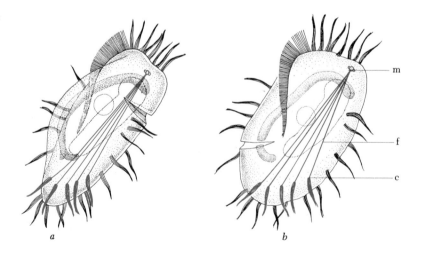

a

b

(and some chemicals) initiates feeding responses. How the ameba captures food depends on what kind of food it is, how active the food is, and whether the ameba is sated or unfed. Amebae exhibit almost no other patterns of behavior; indeed, these appear to be the behavioral limitations of undifferentiated protoplasm.

With the development of organelles, somewhat more complicated reactions may take place. Cilia, flagella, contractile fibrils (myonemes), and specialized sensory areas permit a greater variety of behavior than do all-purpose pseudopodia. For example, directed orientation to light is possible in the green flagellate *Euglena* because it has a light-sensitive eyespot and an opaque spot called a stigma. We can show by passing a shadow over the eyespot and observing the response of the animal that it is indeed sensitive to light. *Euglena* can perform directed (as opposed to random) movements because the eyespot is periodically shaded by the stigma as the transparent organism rotates in swimming, and thus the comparison of intensities (that is, light and no light) necessary for orientation is possible. Protozoa generally cannot perform oriented movements. Stimuli elicit from many protozoa movements whose direction bears no relation to the location of the stimulus. In most flagellates and ciliates, however, any stimulus that is not favorable causes the animal to back away, turn its body aside, and then to resume its forward movement. This avoiding reaction of *Paramecium* is stereotyped; it lacks individuality and originality.

Although protozoa show only a limited number of reactions, the acellular plan of construction does permit a wide latitude of performance. Consider, for example, the differences among species. Some ciliates swim in a monotonous, random fashion; others show a complicated exploratory behavior. *Stichotricha*, for example, lives in a cell of a duckweed (*Lemna*) leaf, where it divides to produce two individuals. One of these extends its front end through an aperture of the cell to feed, while the other, the swarmer, spends most of its early life wandering about inside the cell; eventually it pushes its companion out of the way and emerges through the aperture. If it encounters an abundance of duckweed, it crawls over the leaves and pokes into or enters empty cells. If, on the other hand, there is not much duckweed immediately available, the swarmer makes long swimming excursions until it finds some.

Stalked protozoa such as *Stentor* and *Vorticella* exhibit even more complicated behavior. To a light touch that is continued for a time, the animal may first respond by contracting on its stalk, but it finally becomes indifferent to the stimulus. To repeated touches, it may show a series of reactions. It may first bend, then momentarily reverse its ciliary beat, then contract into its tube, then release its hold and swim away. The order of responses is not fixed, but each tends to remove the animal from the range of the stimulus; that is, all are adaptive. *Stentor* also tends to be selective in its feeding. The hungrier it gets, the less selective it becomes.

The actions of protozoa are characterized not only by responsiveness to external change but also by spontaneous change, that is, by constant unrest, such as the rhythmic action of cilia and the rhythmic contraction of such stalked forms as *Vorticella*. Since the organism is unresting, external stimuli do not operate against a constant background, but against a shifting one. This makes the condition of the animal itself significant in determining how it will react to a given stimulus. Since change leaves its mark, any reactions depend on past events. If *Paramecium*, for instance, has fed on carmine particles for a while, it may later reject them and continue doing so for two to three days. Animals that have been stimulated repeatedly may eventually become indifferent to the stimulus. Stalked protozoa, on the other hand, may give a variety of responses to a continuing stimulus. Here, then, is the first indication that behavior is capable of being modified.

Whether protozoa are capable of learning is still a debated question. It has been reported, for example, that amebae and paramecia are capable of habituation to noxious sensory stimuli such as strong light or mechanical shock, for upon repeated stimulations their responses grow weaker and weaker until in some cases the animals become totally unresponsive. Two criticisms of this work have been offered, however. One is that none of these experiments has satisfactorily demonstrated that such diminished responses lasted long enough to be anything more than adaptations to sensory stimuli. The other is simply that the noxious stimulation may have temporarily injured the organisms, rendering them less capable of responding with each successive stimulation.

To get around these objections, attempts have been made to demonstrate some form of associative conditioning or learning. But each claim has again been met with cogent criticisms. One experiment will serve to illustrate the point. First a sterile platinum wire was lowered into the center of a dish of paramecia. There was no special reaction to it. Next the wire was "baited" with bacteria, and the paramecia responded by congregating around the wire, clinging to it, and feeding. Then, after many such presentations of the "baited" wire, the wire was sterilized and dipped into the same spot, and the paramecia again congregated around it and clung to it. This was claimed to be a well-controlled demonstration of the learning of a new response to the sterile wire. A simple control experiment, however, demonstrated that the training was unnecessary. In this experiment, bacteria were dropped into the dish and the paramecia congregated and fed. Then the sterile platinum wire was lowered for the first time into the same spot where the bacteria had been, and the paramecia clung to it. Careful investigation showed that congregation around the spot the wire touched was due to a residue of bacteria bait. Increased clinging to the wire resulted from the increased acidity the bacteria gave to the medium since further control studies showed that clinging by paramecia is a function of acidity.

The multicellular animal is essentially an interrupted system since the cells of which it is constructed are more or less self-contained units. It is a brick building as opposed to one of poured concrete. Irritability in a multicellular organism thus resides in many discrete units, each capable of different levels of sensitivity and different rates of change. Each may be responsive to different kinds of change in its environment, may require different periods for recovery, and may deliver different kinds of energy by way of response. Generally these units retain a certain independence of action, but they can also frequently affect one another to varying degrees.

Multicellular construction makes possible an almost infinite number of combinations of units with a high level of sensitivity. A state of excitation set up at any one point in a sensitive unit can spread rapidly to all parts of it. When one unit comes into sufficiently close contact with another, the excitation is transmitted across the union. This system of many specialized parts, therefore, can attain a complexity and fluidity that is denied the unitary acellular system. We may call this arrangement of cells, specialized for irritability and conduction, the *nervous system*. Its structural development during the evolution of the animal kingdom has involved the differentiation, shuffling, ordering, and combining of cells.

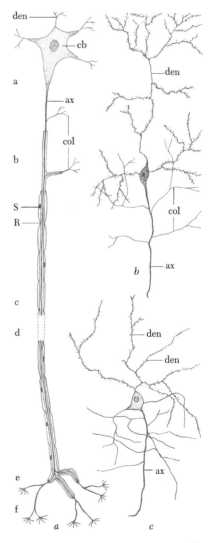

Figure 1.2 (a) *Diagram of a motor neuron, showing cell body,* cb; *dendrites,* den; *axon,* ax; *collaterals,* col; *Schwann's sheath,* S; *node of Ranvier,* R; *area of naked axon,* a; *area of axon invested only with myelin,* b; *area of axon invested with Schwann's sheath and myelin,* c; *broken lines indicating a great extent of axon not shown in this diagram,* d; *area in which the axon is covered with only Schwann's sheath and its nuclei,* e; *area of the naked axon ending in an arborization,* f. (b) A *pyramidal neuron from the cerebral cortex of a rabbit.* (c) Type II *neuron from the cerebral cortex of a cat.* [Redrawn from A. Maximow and W. Bloom, *A Textbook of Histology.* Philadelphia: W. B. Saunders Co., 1941, pp. 169–171.]

The unit of the nervous system is the *neuron*. It comprises the nerve cell body plus all its protoplasmic outgrowths. The average neuron is slightly less than 0.1 mm (millimeter) in diameter, just below the visual range of the naked eye. Although neurons come in a variety of shapes and sizes, they consist typically of three major parts: dendrites, which normally receive excitation and conduct it to the cell body; the cell body, which contains the nucleus; and the axon (nerve fiber), which normally transmits excitation away from the cell body (Figure 1.2). Neurons are highly specialized for conducting impulses, and all nerves are similar in this respect. The neuron is an extraordinarily sensitive cell, and excitation at one point spreads rapidly to all parts, even though the axon may extend a distance of many feet, as, for example, a sensory cell that reaches from the toe to the brain.

That excitation is actually conducted along an axon is amply demonstrated if we pinch (or otherwise stimulate) a fiber whose far end is attached to a muscle. The muscle contracts. Experiments have shown that the passage of such a nerve impulse is accompanied by an electrical change (the action potential) that is a few hundredths or thousandths of a volt in magnitude and a few thousandths of a second in duration. The impulse may travel as fast as 200 mi/hr (in the thickest nerves of man). The nerve impulse is *not* an electric current passing down the nerve; rather it is a complex cycle of electrochemical changes in nerve structure; we shall describe these briefly below.

Since the nerve impulse is the language of behavior, we should know something of its basic properties. These can be studied by placing elec-

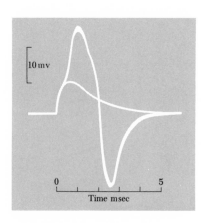

Figure 1.3   *The relationship between the local electrical response (dashed line) to a stimulus and the action potential generated from it, as seen on the screen of a cathode ray oscilloscope attached to two electrodes recording electrical events in a single nerve fiber. The action potential is diphasic because it is being recorded by two electrodes. The local electrical response and the action potential were produced in succession and then were superimposed to show their relationship.* [Redrawn from A. L. Hodgkin, *J. Physiol.*, 90 (1937), 183–232.]

trodes on nerves and observing the changes that occur when the nerves are stimulated. If we place one electrode on the outside of an unstimulated neuron, insert a second inside, and connect the two to an instrument that will record electrical potential, we will see that the outside of the neuron is electrically positive with respect to the inside. The neuron is thus said to be polarized. When it is stimulated (pinched, shocked electrically, covered with chemicals, subjected to drastic temperature changes, and so on), it responds because it is constructed of protoplasm, one of whose basic properties is irritability (responding to change). The response consists of a change in the permeability of the surface membrane, which permits ions to flow into the axon from the surrounding fluid (Figure 1.3).

As a result of this flow, the electrical difference between the inside and outside of the neuron changes. The nerve is then depolarized. If the area or degree of depolarization is small, the nerve uses its energy of metabolism to restore the membrane to its former polarized state, and no nerve impulse is generated. If, on the other hand, the degree of depolarization is great, the nerve cannot restore itself immediately; consequently, electric current flows from the intact areas on either side of the depolarized spot. This flow causes these areas in turn to become depolarized. In other words, the nerve is now stimulating itself. The process continues so that the state of depolarization moves away from the original site in both directions along the nerve. This combination of membrane changes, chemical changes, and electrical changes *together constitute the nerve impulse.* The nerve impulse is usually described, however, in terms of the traveling electrical change (called the *action potential*) since it is the easiest change to measure (Figure 1.3).

The action potential has an all-or-none character. In other words, if it occurs at all, it has the maximum voltage that the axon can produce. Furthermore, it travels along the nerve without any loss of amplitude because it is constantly generated at each point by the neuron. Since the neuron generates the action potential, its characteristics are determined by the nerve and not by the stimulus that began the whole series of events. The situation is roughly analogous to the firing of a gun. The stimulus (the finger on the trigger) produces a pressure (the local excitatory state) that causes the firing pin to release the energy stored in the gun powder and start the bullet on its way. Whether the pressure on the trigger (the stimulus) is fast or slow, or produced by a finger or a hammer, does not affect the flight of the bullet.

A nerve is not equally excitable at all times. After an impulse is generated, the nerve is less sensitive and must recover before another impulse can be generated. Thus, nerve impulses occur as pulses or volleys, even though the stimulus initiating them may be continuous (Figure 1.4).

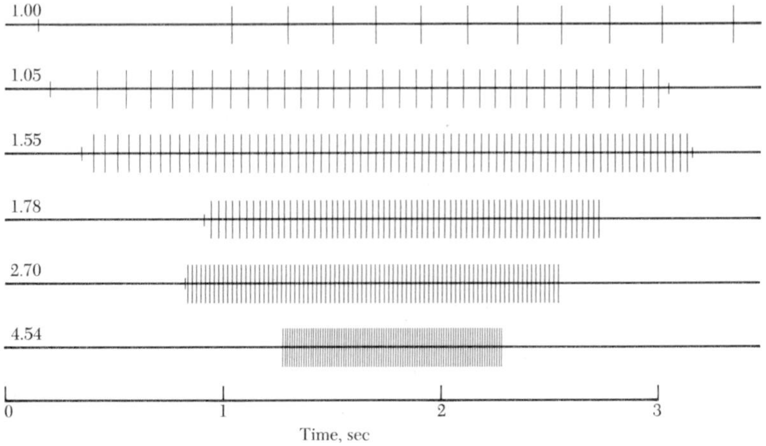

1.00		
1.05		
1.55		
1.78		
2.70		
4.54		

Time, sec

Figure 1.4 The repetitive responses of a single motor fiber from a crab under constant stimulation. The frequency of response increases with the strength of stimulus proportional to the numbers. Note that the amplitude of the action potential does not change. [Redrawn from A. L. Hodgkin, *J. Physiol.*, *107* (1948), 165–181.]

SYNAPSES

We have already pointed out that nerve cells are distinct entities. The places where they come together, where the ends of the axon of one cell come into close contact with the cell body and dendrites of other cells, are called *synapses.* Synapses may be very complex structures, varying in size, location on the nerve cell, and in chemical composition, yielding an enormous assortment of geometrical arrangements. (See Figures 1.5

Figure 1.5 Electron micrograph of rat neocortex at a magnification of 15,000. Presynaptic terminals contain synaptic vesicles and mitochondria; the synaptic cleft at the point of contact with the dendrites contains darkly staining material, presumably serving as a cohesive matrix: d, dendrites; pr, presynaptic terminals; sc, synaptic cleft. [Courtesy of Dr. N. K. Gonatas.]

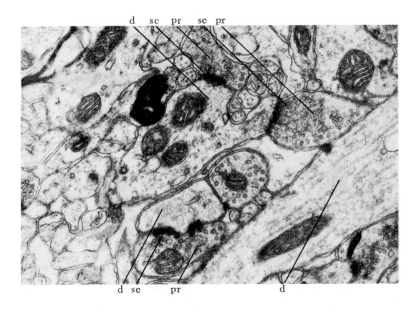

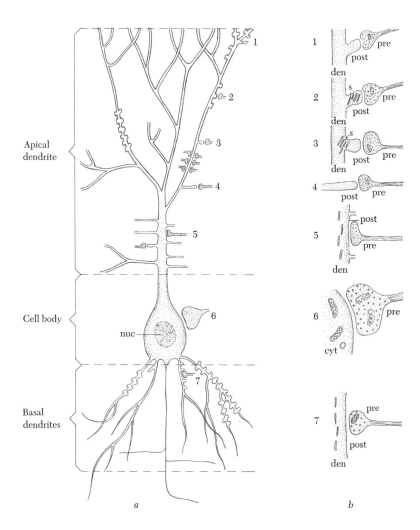

Apical
dendrite

Cell body

nuc

Basal
dendrites

a

1   pre   post   den

2   s   pre   post   den

3   s   post   pre   den

4   post   pre

5   post   pre   den

6   pre   cyt

7   pre   post   den

b

Figure 1.6   *Schematic summary of seven types of synaptic connections (1–7) seen with the electron microscope in the cell bodies and dendrites of nerve cells in the rat's brain. Synapses may vary as to whether the nerve terminals end on cell bodies (5), dendrites (6, 7), or axons (not shown); some terminals end on special dendritic spines (4); some are large, others are small; pre- and postsynaptic areas vary as to the amounts and kinds of intracellular inclusions they contain (as shown schematically in b), suggesting different chemical and physiological properties: cyt, cytoplasm; den, dendrites; nuc, nucleus; post, postsynaptic regions; pre, presynaptic terminal; s, spine apparatus. [From L. H. Hamlyn, "An electron microscope study of pyramidal neurons in the Ammons Horn of the rabbit." J. Anatomy (London), (1963), 97, 189–201.]*

and 1.6.) The terminations of many axons may impinge upon a single cell (Figure 1.7), or a profusely branching axon may impinge upon many cells. Because of the many possible kinds of connections, the presence of synapses in a nervous system introduces great complexities into the pathways over which impulses may travel, just as a multitude of switches in a railroad yard permits more complicated movement of trains than do straight tracks alone. Even greater complexity can occur in the nervous system than in a freight yard, however, because the synapses are not simple switches that all work in the same fashion.

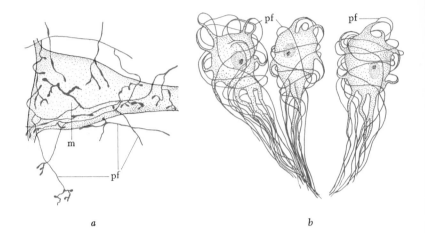

Figure 1.7 **Synapses.** *(a) A motor nerve cell, showing many presynaptic fibers (pf) ending on the same cell.* [Redrawn from L. de No, *J. Neurophysiol., 1* (1938), 194.] *(b) Three neurons from a stellate ganglion, showing the network of presynaptic fibers associated with them.* [Redrawn from de Castro, in *Cytology and Cellular Pathology of the Nervous System,* W. Penfield, ed. New York: Harper & Row, Publishers, Inc., 1932.]

a

b

Furthermore, transmission across synapses occurs usually in one direction only; different synapses delay transmission differently; and synapses eventually fatigue, so that they may fail to transmit for a time.

An even more subtle characteristic of synapses, *summation,* adds still greater complexity to the nervous system. Summation takes place as follows. Excitation crossing the synaptic gap may be insufficient to initiate an impulse in the postsynaptic cell. Repeated excitation of the same strength, however, may succeed; that is, the successive excitations add (temporal summation). Also, excitation from one presynaptic fiber may fail to excite the postsynaptic fiber, whereas simultaneous excitations from many may succeed; they add in space (spatial summation).

The overall effect of summation is called *facilitation;* that is, excitation too weak in itself to cause an effect may facilitate other excitation that also could accomplish nothing by itself. A reverse situation occurs at some synapses where excitation in one fiber blocks synaptic transmission to another (*inhibition*). Inhibition is just as important in nervous coordination as excitation is. In short, as we shall see later, it is the aforementioned functional characteristics and spatial interrelationships of synapses that make integrated action by the nervous system possible.

RECEPTORS

Although some neurons are spontaneously active, discharging without apparent stimulation, most impulses in a neuron arise as a result of stimulation by another neuron or by some change in the environment. The cells that respond to changes in the environment are called *recep-*

tors. Receptors may be free nerve endings of the sort that register pain or they may be nerve cells or fibers associated with very complex accessory structures, such as those that go to make up the eye or the ear. By virtue of its specialized structure, a sense organ is usually sensitive to only one kind of stimulus; the eye *normally* responds to light but will also respond to electrical and mechanical stimulation.

The stimulus, the change in the environment, is some form of energy. Thus, light is detected by means of a pigment that absorbs the energy of photons of certain energy. Heat is derived ultimately from the energy of radiation and transferred by radiation, conduction, or convection. Electricity is detected by the energy of electrons; tastes and smells, by the potential energy existing in the mutual attraction and repulsion of the particles making up atoms; sound, by the energy of moving particles of molecular size. The energy in each case causes a local excitatory state in the receptor, generating an action potential that travels along the connecting nerve. Although the receptor may be specific, the message it sends along the nerve is not. All messages in all nerves are alike in that they are nerve impulses. The stimulus, regardless of its nature, sends only a message that that particular organ was stimulated. If the stimulus was intense, the impulses are close together; if it was weak, the impulses are less frequent (Figure 1.4). The intensity of stimulation, in short, is signaled by the frequency of impulses (each of which fires in an all-or-none fashion) in a fiber and, since different fibers have different sensitivities, by the number of fibers responding. Under continuous stimulation, the sensitivity of a fiber decreases so that the frequency of firing gradually diminishes. This decrease in frequency under constant stimulation is called *adaptation*. The quality of sensation (light, smell, touch) depends on the nature of the connections the sensory nerves make within the nervous system. The relationships among stimulus, neuronal events, and sensation are illustrated in Figure 1.8.

Most animals have mechanoreceptors (responding to mechanical deformation) that are sensitive to stretch, compression, and torque; chemoreceptors, sensitive to tastes and odors; humidity receptors, sensitive to water vapor; and photoreceptors, sensitive to various wavelengths of light. A steady flow of information from mechanoreceptors is necessary for an animal to maintain normal posture, to walk, swim, or fly, to feed, and to manipulate its environment (for example, dig burrows, build nests). Mechanoreceptors are also important in supplying information for the proper working of many internal organs. Hearing, a form of mechanoreception, is important in many animals for courtship, advertising their territory, warning, and orientation by echolocation (as in bats and porpoises). Detection of tastes and odors is important in all feeding

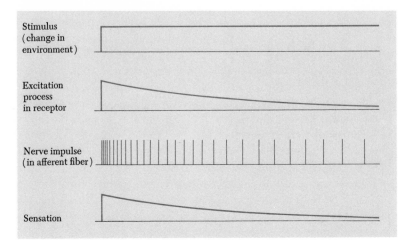

Figure 1.8 *Diagram showing the relationships among the stimulus, the local excitatory state, the action potential, and sensation.* [Redrawn from E. D. Adrian, *The Basis of Sensation.* London: Christophers, 1949.]

activities, many sexual and reproductive activities, social activities, and forms of orientation that involve trail following.

Clearly, the nervous system does not sit idly by in a state of passive rest. Because of the spontaneous activity of many neurons, it is in a state of continuous background activity. Messages coming in from the sense organs in contact with the external and internal environments may be thought of as constantly modulating this background activity. It is the interaction of background activity and messages from sense organs that produces behavior.

IN THE SIMPLEST NERVOUS SYSTEMS KNOWN, NEURONS AND SYNAPSES ARE already fully developed. The physiology of neurons is probably the same in all animals, and the range of variations among synapses is probably as great in any particular animal as between widely unrelated species. Within the coelenterates evolutionary changes involve primarily new grouping and arrangement of neurons. These developments permit greater complexity of integration. Almost from the beginning, two major trends have been evident in the evolving nervous system: one toward a division of labor, in which different neurous or their parts become specialized for different jobs, the other toward segregation, in which like units become grouped together. The overall effect of this grouping has been a tendency toward centralization.

Among the simplest nervous systems known are those in the hydras, jellyfish, sea anemones, and corals (Coelenterata). Even within this group there is variation from a condition of uniformity and complete

THE NERVOUS SYSTEM OF HYDRA

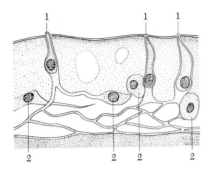

Figure 2.1 *Sensory cells (1) and cells of the epidermal nerve net of Hydra (2).* [Redrawn from L. H. Hyman, *The Invertebrates.* New York: Mc-Graw-Hill Book Company, 1940, Vol. 1. After J. Hadzin, 1909.]

Figure 2.2 *An epitheliomuscular cell in the epidermis of Hydra, showing supporting fibrils (1) and myoneme (2) (contractile fibril) of muscular base.* [Redrawn from L. H. Hyman, *The Invertebrates.* New York: Mc-Graw-Hill Book Company, 1940, Vol. 1. After G. Gelei, 1924.]

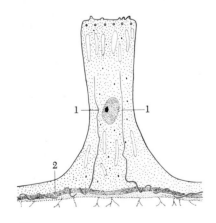

diffusion to one in which there is a considerable division of labor and almost the maximum degree of centralization permitted by a radial plan of body construction.

The nervous system is seen in its simplest form in *Hydra*, where it consists of three functional divisions: cells that sense changes in the environment (receptors), cells that conduct excitation to various parts of the body, and cells that respond to these changes by movement (effectors). Not all the surface of *Hydra* is equally sensitive to changes in the environment. Certain epithelial cells (receptors) that occur singly or in patches on the external surface and on the internal surface that is in contact with food are especially sensitive to mechanical and chemical changes (Figure 2.1). Other sense cells (epitheliomuscular cells) have flattened bases equipped with contractile fibers (Figures 2.2). Whereas the simple sense cells transmit excitation through their basal fibers to other cells, the epitheliomuscular cells transform the energy of the stimulus into work; that is, their bases contract. They are thus said to be independent effectors. *Hydra* and other coelenterates possess, in addition, nonnervous, independent effectors that are of great importance in behavior. These cells (nematocysts) consist of small harpoonlike structures that are capable of being discharged when they are properly stimulated.

The simple sense cells transmit their excitation either to the second element of the nervous system, the conducting elements, or directly to the muscles. Aside from independent effectors, a direct connection between sense cells and muscle is the simplest possible association between sensing element and working unit. It allows very little flexibility of action. On the other hand, when the sense cell ties into the conducting system, far more varied behavior is possible because of the increased number and complexity of connections with effectors.

The primitive conducting system is a nerve net. Considerable mystery still surrounds this system in *Hydra*. Although the electron microscope shows no neurons, silver and methylene blue stains reveal neuronlike cells. They are unipolar, bipolar, and multipolar. They conduct slow action potentials (2 to 50 msec). In some coelenterates the net is believed to be continuous and unbroken (syncytium), but in *Hydra* it is believed that the cells never lose their identity even though they come extraordinarily close together. This arrangement appears to be a synaptic system yet it does not exhibit all the characteristics of one. For instance, it does not possess one-way transmission nor does it transmit without loss in strength. One portion of the network lies close to the external surface of the animal (as does the conducting system of protozoa); the other portion lines the gastric cavity. All parts of the animal are in slow communication with one another through the mazelike web. Since cell bodies and processes are about equally distributed through the organism, one part of the system is essentially like another. But it was probably a system like this, which in *Hydra* forms an indiscriminate communications network, that gave rise to the central nervous system of higher animals.

The third important element in the behavior of *Hydra* is the muscular system, which is composed of the dual-function epitheliomuscular cells already mentioned and single-celled contractile units that are greatly extended and branched and so form thin muscular sheets. These cells, like all muscle cells, are contractile units with a high level of irritability. For work to be done, the irritability must be altered in some way to set off the contractile mechanism. The function of the nervous system is to affect the irritability.

Sea anemones look like larger, more complicated, rococo *Hydra* and are essentially that from a nervous and muscular point of view. Structurally the nervous system is not appreciably more complex. It is, however, more synaptic in nature; that is, the neurons are not only more clearly separated but may be functionally more discrete. They also consist primarily of bipolar cells.

Characteristically, sea anemones (and other polyps and medusae) possess two or three discrete nerve nets occupying identical regions. One type is a wide-mesh lattice consisting of large bipolar cells. The fibers may run in parallel courses, especially in the mesenteries. These bipolar cells are associated with fast muscle. Some of the cells function as motor neurons; others are interneurons that permit through conduction at high velocity. This net is therefore designed for through conduction and rapid distribution of messages. The other common type of net consists of small multipolar cells; this type conducts slowly and diffusely (Figure 2.3).

In those species which lack through-conducting systems, responses to external stimulation tend to be local rather than symmetrical and coordinated. *Calliactis* and *Metridium*, on the other hand, possess well-developed through-conducting systems and, as might be expected, exhibit symmetrical responses to external stimulation (see Figure 2.4 on page 21).

The greatest advance, however, lies in the effector system. Instead of being isolated fibers, the muscles are organized into circular and longitudinal muscle cylinders. The arrangement of muscles and mesenteries (supporting tissues) is such that these animals tend to be bilaterally rather than radially symmetrical.

Figure 2.3 **Two nerve nets in the aboral ectoderm of Velella as revealed by silver staining.** *The large, thick fibers probably represent a continuous net (anastomosis); the thin fibers, a net in which the cells are discontinuous.* [From G. O. Mackie, *Quart. J. Microscop. Sci. 101* (1960), 119–131.]

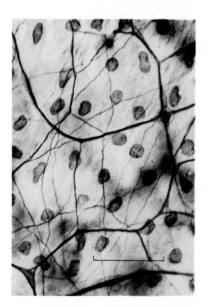

Colonial coelenterates are remarkable in that they behave at one and the same time as groups of individuals and as one organism. The degree to which a colonial nervous system is developed varies among species. Some species have no such system and hence no colonial coordination. Some may have two colonial conducting systems—a slow, local, diffuse system, and a fast, through-conducting system. In addition, each individual of the colony has its own nervous system. The behavior varies according to the kind of system present.

In animals that are radially symmetrical, any development of the nervous system toward centralization clearly will be molded by the basic symmetry of the body and, as a moment's thought will reveal, will also be constrained by that symmetry. The maximum degree of centralization that can be achieved is seen in the hydroid medusae (jellyfish) and in starfishes and their relatives.

The nerve network in the jellyfish is channeled into two nerve rings in the bell, or umbrella, an arrangement made possible when some processes of the neurons became longer than others and were grouped into parallel bundles (Figure 2.5). Such an arrangement permits speedy directional conduction rather than slow diffusion. The large upper ring receives fibers from the sense organs in the margin of the bell and also supplies the musculature that activates the bell. The two rings are connected by a network of neurons (plexus).

The first true sense organs in the invertebrates are found in the jellyfish. One kind (ocelli) responds to light; another kind (statocysts) responds to changes in position. The ocelli (Figure 2.6) are patches of pigment cells that may be cup-shaped and equipped with a lenslike body, and are interspersed with nerve cells. The statocysts (Figure 2.7) are groups of sensory cells associated with a round concretion of organic material and calcium carbonate.

The structures described in the previous paragraphs are the sensing and responding systems with which the coelenterates have to work in coping with their environment. Most of their behavior is concerned with feeding, locomotion, and protection from noxious stimuli. The behavior varies in complexity from rather simple patterns in the small stalked

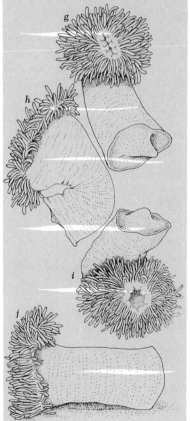

Figure 2.4   *The swimming sequence in a sea anemone (*Stomphia coccinea*).*
(a) Usual position. (b) Response to stimulation by contact with a starfish (a
predator). (c) Extension of column by circular muscles after contraction of
longitudinal muscles. (d–f) Lateral bending by parietobasilar muscles. (g–i)
Swimming. (j) Inactivity. [From P. N. Sund, *Quart. J. Microscop. Sci.,* 99
(1958), 401–420.]

Figure 2.5   *Fine structure of the nerve ring of the jellyfish* Gonionemus:
(1) upper nerve; (2) fibers that are crossing to (3) lower nerve; (4) connecting
fibrils to subumbrella net. [Redrawn from L. H. Hyman, *The Invertebrates.*
New York: McGraw-Hill Book Company, 1940, Vol. 1. After Phyde, 1902.]

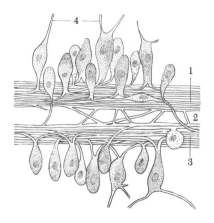

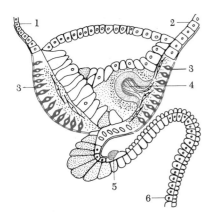

Figure 2.6 *The complex ocellus of a jellyfish: (1) exumbrella epidermis; (2) subumbrella epidermis; (3) sensory epithelium; (4) pigment cup of ocellus; (5) statocyst; (6) velum.* [Redrawn from L. H. Hyman, *The Invertebrates.* New York: McGraw-Hill Book Company, 1940, Vol. 1. After A. Linko, 1900.]

Figure 2.7 *Closed type of statocyst that is found in the jellyfish Obelia: (1) statolith; (2) sensory cells.* [Redrawn from L. H. Hyman, *The Invertebrates.* New York: McGraw-Hill Book Company, 1940, Vol. 1. After O. Hertwig and R. Hertwig, 1878.]

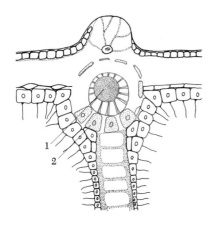

species like *Hydra* through more complicated movements in the stalked sea anemones, culminating in the complex behavior of such free-swimming medusae as *Gonionemus* and colonies such as the Portuguese man-of-war. In general, all responses are slow and stereotyped, and isolated fragments of the animal respond as well as the whole animal.

In the presence of adverse stimuli, the general behavior pattern is one of slow contraction of all or part of the body, depending on the point of contact of the stimulus and its intensity. The mechanism has been carefully studied in the case of protective closure of the oral disc of the sea anemone (*Calliactis parasitica*). Closure is obviously a response that protects the delicate disc and tentacles from injury. The sensitivity of the sphincter muscles is low and adaptation is rapid. Messages are transmitted fairly rapidly from the site of stimulation through the nerve net. Speed is possible because the organism has a through-conducting system—that is, certain neural pathways conduct in an all-or-none manner without hindrance. Somewhere in the system facilitation occurs—one nerve impulse does not normally cause the muscles to contract; bursts of impulses are required. The low sensitivity, rapid adaptation, and faciliation are all characteristics that insure response only when absolutely required.

Feeding behavior may be even more complicated than the mechanism just described. In *Hydra* it involves these steps: The tentacles grasp solid objects, the nematocysts discharge, the objects are conveyed to the mouth, the mouth opens to receive the food, the tentacles relax after the food has reached the mouth, and the rim of the mouth glides over the food to encompass it, after which the mouth closes. The distal part of the column then undergoes circular contraction, forcing the food into the middle region, where it is retained from one-half to several hours. Eventually indigestible material is rejected. The fine coordination and the anticipatory opening of the mouth seem to be impossible feats for a simple net nervous system. We still do not understand the mechanism of feeding, although careful studies of nematocysts have shed some light on the problem. There are four types of nematocysts in *Hydra,* each with a different function. In two types the threshold is lowered (sensitivity is increased) by chemicals diffusing from the food, so that prey bumping into the tentacles elicits an immediate response. One of these two types of nematocyst entangles prey and holds it; the second, discharging later, penetrates and kills the prey. In the third type, chemicals from food raise the threshold to touch so that firing does not occur in the presence of food. These nematocysts are employed in locomotion. Accordingly, food inhibits locomotion. A fourth kind of nematocyst has its threshold to touch lowered by diffusible substances from noxious animals and thus is employed in defense.

It was formerly believed that *Hydra* was restricted to living prey from Nemathelminthes or higher phyla and that the feeding reaction depended exclusively on stimulation by glutathione released from wounded prey. These conclusions have not been substantiated. *Hydra* feeds on many things, including filter paper soaked in beef broth, protozoa, algae, and mud.*

In *Metridium*, feeding behavior is more than just a simple succession of responses to stimuli. As in *Hydra*, it consists of a preparatory behavior, a discharge of nematocysts, and movements conveying food to the mouth. The presence of food juices in the vicinity often induces a characteristic preparatory movement arising from the alternate contractions of the longitudinal and circular muscles. The disc expands, the stalk elongates, and the animal may make swaying movements. All these movements increase the animal's chances of coming into contact with food. Each element of the pattern involves the coordinated activity of several muscle systems, which have long latency and smooth contraction. Elongation involves the reciprocal inhibition and successive activation of two antagonistic muscle systems, the circular muscles and the longitudinal muscles. The final act involves the reflex movements that carry food to the mouth.

The preparatory feeding movements of *Metridium* are not a unique pattern brought about by stimulation; they are a modification of movements that are going on in the animal all the time, even in the absence of stimulation. Sea anemones in a constant environment show rhythmic spontaneous activity which is, however, so slow that it can be observed only by the use of time-lapse photography. This activity is produced by the alternate contraction of the longitudinal and circular muscles, the same ones employed in the preparatory movements. In *Hydra*, spontaneous movements are few, but contractions and expansions do occur at intervals without any apparent cause, and the tentacles may move about freely.

Spontaneous activity is the basis of the swimming movements in such jellyfish as *Gonionemus*, where by vigorous contraction of the ventral surface of the animal, water is forced rhythmically out of the bell. Experiments have shown that the origin of activity is the nerve ring, for when the ring is cut, the coordination of the movements of the bell is destroyed, although small pieces of the ring and attached bell continue to contract rhythmically after separation from the animal. Ingenious cutting experiments with jellyfish and with sea anemones demonstrate strikingly that, in the latter, normal contraction of the animal is not hindered. These results indicate that one nerve net of anemones

* For a complete discussion, see *Biol. Bull.*, *122* (1962), 343.

conducts in a diffuse manner in many directions, whereas in the jellyfish the wave of excitation originates in the marginal bodies, those cell concentrations consisting of a statocyst and grouped neurons. If all marginal bodies but one are removed, a cut can be made in such a way that the wave of excitation will be trapped. Such a wave courses around the bell in a circular path for as long as 11 days, during which it travels over 450 mi. Such experiments show not only that conduction is channeled but also that at this evolutionary stage, the nervous system has groups of neurons that have endogenous rhythmic activity and act as pacemakers.

The spontaneous movements exhibited by coelenterates are by no means constant. They vary with the condition of the animal. *Hydra,* for example, when deprived of food for some time extends its stalk, increases the movements of its tentacles, and eventually moves from its place of attachment. It moves by any one of three methods: by basal gliding (a mechanism not clearly understood), by pulling itself along by the tentacles, or by looping and somersaulting. *Gonionemus,* when deprived of food, "fishes" by swimming to the surface of the water, turning upside down, and then floating slowly downward with the tentacles widely extended. It may do this for hours.

The behavior of these animals is clearly being modified by internal conditions. When coelenterates are food-sated, the nematocysts will no longer discharge. When animals have been stimulated repeatedly in such a way as to cause contraction, they eventually cease to contract. In these and other instances, the results can be satisfactorily explained by sensory adaptation. There is no unequivocal evidence of learning of a higher order, however.

Here, then, are behavior patterns that are simple, stereotyped, slow, and sometimes representative of modifications of an endogenous activity of the simple nervous system. Complexity of behavior is clearly related to the development of the musculature, the degree to which the nervous system departs from a simple nerve net, and the degree of specialization of sense organs and independent effectors. Where the net is undifferentiated and the muscles isolated, there are few reflexes, and a part of the animal is nearly as efficient as the whole. Where the muscular system has become complicated and the nervous system concentrated into through-conduction pathways, as in the swimming medusae, the action of the individual parts has become coordinated. The coordination stems primarily from the topographical arrangement of nerves and the number and kinds of nets. In the coelenterates we also see for the first time an example of the degree of nonnervous coordination (the nematocysts) that can be achieved by differential response to various chemicals in the medium.

COELENTERATES DID NOT EXHAUST THE POTENTIALITIES OF A RADIAL nervous system, even when they abandoned sessile life for freedom of motion. However, the ultimate in radial nervous systems was achieved by a different and unrelated group of animals, the echinoderms (sea cucumbers, sea urchins, and starfish). Neither tentacular nor umbrellar movement had given the nervous system much latitude for change. With the development of tube feet in sea urchins and starfish and with the added freedom of movement permitted by arms in the starfish (especially the brittle stars), coincident with a more diversified nervous system, behavior achieved considerable versatility. One of the great advances of the nervous system was the increase and ordering of the neurons between the sense organs and effectors. As the nerve net condensed into nerve tracts and moved away from its location near the surface of the animal, the distances between receptors and effectors became greater. Direct synapses between receptors and effectors disappeared as additional neurons came into being to make connections. These intermediate neurons, the internuncials or associational neurons, made possible a variety of connections between sensory (afferent) neurons and motor (efferent) neurons. By providing more synapses and alternate pathways, they enhanced greatly the capabilities of the nervous system.

Various stages of the evolutionary developments just outlined are preserved in the echinoderms. Sea cucumbers have not advanced far. They have a nerve ring surrounding the buccal (mouth) cavity, nerves to the tentacles (when tentacles are present), five radial nerves supplying muscle fibers of the body wall, and a general body plexus. This system is consonant with the uninspired behavior exhibited by these sluggish animals, whose repertory consists essentially of moving the tentacles for food, righting themselves, and burrowing in the ocean bottom.

THE STARFISH NERVOUS SYSTEM

Figure 3.1 **Diagrammatic representation of the nervous system of the starfish.** *Excitation from plexus is transmitted via the nerve cord to motor neurons (mn), then to neurons (n), then via three main tracts (1, 2, 3); str, stroma; oss, ossicle; pl, plexus.* [Redrawn from J. E. Smith, "Physiological Mechanisms in Animal Behaviour," in *Symposia Soc. Exp. Biol., 4.* New York: Academic Press, Inc., 1950.]

The basic plan of the nervous system in starfish is similar to that of sea cucumbers, consisting of a circumoral ring, a radial nerve to each of the five arms, and a dermal plexus. In detailed anatomy, however, it is wondrously complex, being differentiated not only into sense cells and motor neurons, but also into associational neurons, nerve tracts, and reflex arcs of various degrees of intricacy. Many activities of starfish can be understood in terms of these units. Their interrelationships illustrate strikingly how behavior is a function of the nervous system.

The outer surface of a starfish is equipped with spines and pincerlike structures (pedicellariae), which can hold and paralyze small animals that come into contact with them. The outer ectoderm consists of epithelial cells, mucus glands, and about 4,000 sensory cells per square millimeter of surface. Lying beneath these cells is a nerve plexus, the outer layer of which consists of randomly arranged fibers in synaptic connection with the sensory epithelium. Still deeper, the plexus is organized into linear tracts. All of this structure lies outside the thick, fibrous stroma that constitutes the main bulk of the body wall. Where thin spots exist in the body wall between the ossicles, synaptic connection may occur with the motor part of the system that lies inside the stroma. The inner surface of the stroma—that is, the part facing the body cavity—is lined with epithelium where the muscles and motor fibers are located. Thus, the sensory system and associational pathways of the epidermis lie outside the stroma and the motor tracts lie inside.

The calcified plates of the arm are so situated that the arm, its associated nerve arcs and motor centers, and the tube feet are arranged segmentally (Figure 3.1). Axons of the motor neurons in the radial nerve cord encircle the tube feet. They join a lateral motor center, the

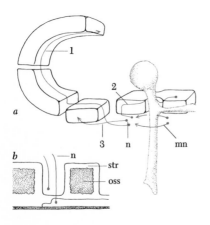

axons of which extend in three tracts to various parts of the arm. Here they make synaptic connections that eventually reach the muscles.

The contributing roles of these differently organized parts of the nervous system are well illustrated by the different response patterns that are elicited when one spot on the dorsal side of one of the arms is lightly pressed. First, the spines bend toward the spot and the pedicellariae open and close their pincers. This response is extremely localized. It is certainly mediated through the dorsal plexus. In its action, it resembles the nerve net of coelenterates. The foot immediately below the spot that was stimulated is then activated. This response is regulated via the segmental arcs in which transmission is through-conducting and polarized (that is, in one direction only). Next, adjacent feet respond as a result of extrasegmental transmission via the radial cord. Finally, excitation involves the radial cords and nerve ring and coordinated movement follows. Thus we see two aspects of nervous control: one peripheral and reflex, and the other central and generalized.

Much of the behavior of starfish is based on activities of the tube feet. In addition to being protracted and retracted, the tube feet make postural movements associated with stepping. Protraction and retraction are unoriented, wholly reflex responses. When a region near a foot is stimulated, the foot retracts; when a wave washes over the animal, all feet retract, not as a result of coordination, but as independent members all reacting to a common stimulus. Protraction and retraction are antagonistic movements that reflect reciprocal states of excitation and inhibition in motor neurons of the foot. Neurons mediating postural movements, however, reflect changing states of central excitation.

Obviously, if there were not some coordination of stepping activity, the starfish would never progress. The coordination of feet within a given arm is effected by a control center at the base of the arm. An arm whose feet are stepping in a distal direction is dominant and imposes its direction on the others. Excitation to all feet of the animal is thus through-conducted from the dominant center via circumoral and radial nerve tracts. In some species, there is a tendency for the number 2 arm to lead most frequently. In all probability, some intrinsic feature of nervous organization, even possibly some retention of traces of bilaterality from larval days, is the basis of this behavior.

Generally, there are autonomous transfers of dominance from one foot to another. This suggests that there are periodic rises and falls of activity in the different centers. It is even possible that this might explain why starfish, given the problem of removing a rubber tube placed on one arm, may "solve" the problem in a variety of ways. These various levels of nervous activity and the waxing and waning of excitation may also explain how a starfish can right itself in a number of ways rather than merely by a single stereotyped method. It can, for example, right itself by somersaulting, by folding over, or by assuming a tulip form that causes it to flop over in an upright position.

In addition to righting themselves, echinoderms perform a number of other behavior patterns. Many starfish and sea urchins, for instance, exhibit a genuine brooding of eggs. In some cases the mother arches the disc so that it becomes concave on the oral side. The arms are pointed ventrally so that their bases form, with the concavity of the disc, a space in which eggs develop to the stage of tiny stars. Some sea urchins carry the young on the body. Many species characteristically burrow. Some bore into solid rock when they are small, enlarge the bottom of the hole as they grow, and live there permanently, being unable to get back out through the small hole made in their youth.

In relation to feeding, a variety of behavior patterns have been observed; the most common is that of the starfish that opens bivalve shellfish and digests the contents by everting its stomach. Some species of echinoderms are able to detect food from a distance. For example, one species of sea urchin, which lives in the sand in a burrow whose walls are plastered with mucus, keeps a pair of extensile appendages thrust through the surface hole and with widely opened terminal processes explores the sand surface, picking up food particles by way of adhesive secretion. The appendages then retract and deliver the food particles to the region of the mouth.

Some echinoderms also show defense reactions. Thus, sea urchins have a shadow reflex whereby spines are erected toward the source of danger. If the stimulus continues, the animal changes its behavior and retreats. The shadow reflex may be given by an isolated piece of shell. It is mediated by radial nerves, the nerve ring not being necessary.

Temporary modifications in behavior have been demonstrated in a number of different situations, but it has never been possible to rule out sensory adaptation, injury, or direct physical change as explanations of these modifications. For example, in one experiment with starfish, the animals were prevented from using their dominant rays in turning over by being restrained with a glass rod. The starfish used other rays to turn, and after 18 days of such treatment, 10 times a day, the nondominant rays were used in turning over even when the glass rod was not applied. It was subsequently shown, however, that this same result could be obtained without the training procedure simply by irritating the dominant rays with mild acid or by rubbing them with the glass rod. After one or two such treatments, turning over was accomplished with the nondominant rays. Whether a starfish can learn to use any particular arm in righting itself or can profit by experience in escaping, for instance, from pegs inserted in angles between the arms is questionable. On the other hand, a clear case of learning in the Pacific starfish (*Piaster giganteus*) was demonstrated under more natural conditions. Captive specimens that habitually rested on the walls of a tank kept in darkness learned in four trials to associate a light stimulus with food. When a light was turned on, the starfish descended to the bottom of the tank, a distance of 12 in., where food would be presented. Associating light with food has relevance to natural conditions. A situation has been described in which free-living specimens associated the shadow of a pier with food and learned to remain under it. The food consisted of clumps of mussels that grew in great abundance on the steel pilings of the pier and fell from time to time to the bottom immediately below.

SUMMARY

In our perusal of the animal kingdom thus far, we have observed a tendency toward the formation of specialized cells for selective response to environmental changes (receptors), the elaboration of these into groups associated with other tissues (sense organs), the development of specialized response systems (muscle cells and organized muscle sheets), and the channeling of conduction of excitation (nerve nets and nerve tracts). Specialization of function has gone hand in hand with a trend toward spatial separation. Sensory systems obviously tend to remain near the surface of the animal; response systems lie at deeper levels; consequently, the sensory and motor portions of the nervous system localize.

Conducting elements, which began as nerve networks lying close to the surface, in evolving progress inward into the animal and become the connecting link between receptors and effectors. The connections have been rather simple and direct. The number of neurons intervening between receptor and effector, however, seems to be increasing. These interneurons (internuncials) are, on the whole, not localized but are scattered in, and make up, the fiber tracts. The fiber tracts, which represent a condensation of the nerve network, are still made up of a mixture of fibers and their cell bodies.

These developments have permitted animals to make more refined assessments of environmental changes, more diversified movements, and more rapid, directed, and coordinated responses. Nevertheless, the coelenterates and echinoderms remain animals with limited responses, a low level of coordination, and an absence of central control. Centralization of the nervous system is severely restricted by the radially symmetrical body plan.

4 BILATERAL NERVOUS SYSTEMS: WORMS AND MOLLUSCS

WITH THE ASSUMPTION OF A BILATERAL BODY PLAN, MOST ANIMALS acquired a longitudinal axis and a definite front and back. In this setting, the nervous system began a vital course of development whose end is not in sight and whose behavioral potentialities seem limitless. Such evolutionary trends as a functional division of the nervous system, condensation of the nerve net into more direct conductive pathways, and the interpolation of internuncials between sensory and motor systems as these moved farther apart—are perfected in bilaterally symmetrical animals. The refinement of sense organs, culminating in the exquisitely attuned olfactory receptors of insects, the vertebratelike eye of the octopus, and the beautifully efficient compound eye of the dragonfly, continues in various branches of the animal kingdom. The problem of connecting nerves with muscles (neuromuscular mechanism) had been solved in a most efficient manner very early in evolution; hence, no further major changes took place. The effector system itself, the muscles, became more and more complex as animals developed articulated skeletons (external in the invertebrates, internal in vertebrates) that permitted a versatility of movement heretofore unseen.

The truly great advance that bilaterality permitted was centralization of control. In simple animals, individual parts are rugged individualists; in complex animals, the parts have their actions subordinated to the activity of the whole. In the early stages of the evolutionary development of the nervous system, condensation of neural elements into centers occurred simultaneously in various parts of the body, but eventually the anterior end of the animal assumed more and more control as other centers became subordinated to it or lost. Although the net plan of innervation lost ground to a more centralized system, it was never completely abandoned. It is retained in some form for special purposes in almost all animals. Although it is essential for normal locomotion in echinoderms and lower chordates (for example, balanoglossids), it is less important in annelids and molluscs. In the earthworm, the net system is probably a sensory relay system. In the mammalian intestine, a net is primarily responsible for coordination of peristaltic movement. Throughout the animal kingdom, then, where sluggish movements are called for, a diffuse, slowly conducting net is efficient.

Many of the stages of the evolutionary development of a bilateral nervous system are preserved in the existing animal forms, ranging from flatworms to arthropods. The beginnings of the system are to be found in the simplest flatworms (Acoela) and primitive molluscs (chitons), where it is still a network located near the surface of the body. In some

Figure 4.1   (a) Part of the submuscular plexus of a flatworm (Acoela), showing the statocyst (1) and the brain (2). [Redrawn from L. H. Hyman, The Invertebrates, Vol. 2. New York: McGraw-Hill Book Company, 1951.] (b), The nervous system of Chiton: bc, buccal commissure and ganglia; cc, cerebral commissure; pvc, pallovisceral commissure; pn, pedal ganglion and nerve; src, subradula commissure. [Redrawn from L. A. Borradaile and F. A. Potts, Invertebrata. New York: Crowell Collier and Macmillan, Inc. 1935.]

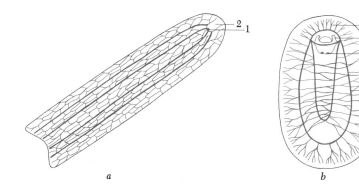

a   b

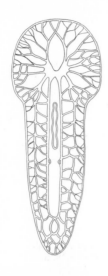

Figure 4.2 **Part of the nervous system of a polyclad flatworm.** [Redrawn from L. H. Hyman, *The Invertebrates.* New York: McGraw-Hill Book Company, 1951, Vol. 2.]

species, though, there is a faint suggestion of an anterior concentration of nervous tissue and a tendency toward stronger development on the longitudinal strands of the net than of strands running in other directions (Figure 4.1*a*, and *b*). The cells making up the strands have repressed the more or less equal development of all their extensions and have greatly elongated one process (the axon). Some specialization of sense organs, notably the statocyst and "eyes" in the form of pigment spots or cups, has occurred in the Acoela. The effector system is still essentially that of a longitudinal and a circular sheet of muscle cells, a "muscle field system." A nerve network can adequately control such a diffuse effector system; however, diffuse nervous control and the lack of mechanical versatility of a muscle field preclude the execution of complex movements. As a consequence, complicated and rapid behavior is denied these animals.

In polyclad flatworms, the nervous system has retreated from its exposed position at the surface of the body to form a submuscular plexus. Longitudinal cords are emphasized, the ventral ones becoming strongly developed (Figure 4.2). These nerve cords, like those of Acoela and chitons, are not pure fiber tracts; they are mixtures of cell bodies and axons. Accordingly, one part of the cord, and hence of the animal, is practically as talented as another. This is true even though there is a tendency for the cords to thicken at the anterior end. In the polyclad flatworms, the "brain" attains great complexity (Figure 4.3), but the word "brain" is misleading here in that it implies too much behaviorally. The so-called brains are concentrations of cells employed principally to funnel into the nervous system the increased number of fibers arising from the greater multitude of sense organs that are now concentrated at the anterior end of the animal (Figure 4.4).

Like the marginal bodies of coelenterates, the brain initiates activity that passes down the cords and, in this sense, influences locomotion. As

Figure 4.3 **Brain of a polyclad flatworm, showing the nerves (a) and various ganglion cells (1) and fiber tracts (2–6) within the central mass (b).** [Redrawn from L. H. Hyman, *The Invertebrates.* New York: McGraw-Hill Book Company, 1951, Vol. 2. After D. Hadenfeldt, 1929.]

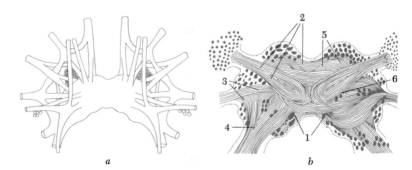

a

b

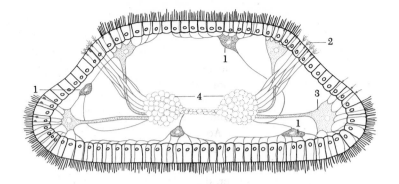

Figure 4.4 *Diagrammatic section through the head of a flatworm showing the brain (4) and the various kinds of receptors: (1) touch receptor; (2) chemoreceptors; (3) rheoreceptors.* [Redrawn from L. H. Hyman, *The Invertebrates.* New York: McGraw-Hill Book Company, 1951, Vol. 2. After G. Gelei, 1930.]

long as a piece of nerve cord is left intact in a fragment of planaria, however, spontaneous movement and some coordinated responses persist. Decapitated flatworms tend to be inactive, not entirely because loss of the anterior ganglion impairs the functioning of the rest of the system, but also because the animal is deprived of the sensory information that initiates and regulates much of its behavior. These animals are still very much stimulus-response systems. Their responses to light, currents, and chemicals are stereotyped. Greater accuracy in orienting to the source of a stimulus is exhibited by flatworms than by lower animals because a greater development of sense organs permits comparisons of intensities of a stimulus successively (klinotaxis) or simultaneously (tropotaxis).

Habituation, the tendency to ignore eventually a stimulus that produces no harmful effects, is common among flatworms. There is also evidence that they can be conditioned to light and to mechanical stimuli and that the anterior end of the animal is not essential for this accomplishment. In other words, the nascent bilateral nervous system, like the radial system, acts not only in accordance with changes impressed upon it at the moment, but is also influenced by what has happened before.

It is clear that the mere concentration of nervous tissue at the front end is insufficient for the performance of complicated behavior. The pioneer possessors of bilateral systems do not outshine in behavior the most highly developed coelenterates nor even some of the protozoa. The rotifers, for example, which resemble ciliate protozoa remarkably in size and superficial structure, do not—despite having a brain, nerve cords, sensory and motor nerves, and a bewildering array of muscles (Figure 4.5)—appreciably surpass the ciliates in behavior. Nor do the Entoprocta, small sessile metazoans that have nerves, a main nerve

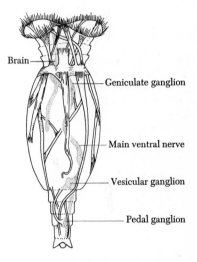

Brain

Geniculate ganglion

Main ventral nerve

Vesicular ganglion

Pedal ganglion

Figure 4.5 *The musculature (shown in heavy shading) and nervous system of a rotifer.* [Redrawn from L. H. Hyman, *The Invertebrates.* New York: McGraw-Hill Book Company, 1951, Vol. 3. After E. Martini, 1912.]

BILATERAL NERVOUS SYSTEMS

mass, and muscles, excel the stalked protozoa they so closely resemble. It is the cellular organization within the central mass plus a versatile mechanical system to play upon that counts.

In the nerve net, it will be recalled, the cells and their processes are more or less evenly scattered in space. As certain processes elongated to the exclusion of others, the bundles of fibers so formed still retained cells within their confines (Figure 2.3), but as animals evolved, there was an increasing tendency to reserve cords for conducting elements and to gather the cell bodies together into localized masses (ganglia) (Figure 4.6.) Cords, being through-conducting systems, transmit faster than does a net. The ultimate in speed was achieved by the development of giant fibers, which transmit faster because of their great diameter. These have fewer synaptic connections, and hence fewer delays, than do ordinary fibers. These express lines of the central nervous system are found in annelids, molluscs, and arthropods, where they are used in such escape mechanisms as the violent tail-snapping response of lobsters. With these developments, the central nervous system sank deeper into the body, where it became protected with various supporting tissues. In these relatively deep locations, more or less equidistant from sense organs and muscles, it became the center where fibers to and from different parts of the body meet. It is capable of many kinds of neural activity, as, for example, reciprocal excitation of antagonistic muscles, intersegmental and chain reflexes, facilitation, central and peripheral inhibition, rapid conduction, slow conduction, wave progression with and against the direction of locomotion, central excitatory states.

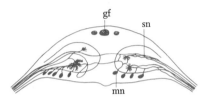

Figure 4.6 **Relationships among the terminations of sensory neurons (sn) and motor neurons (mn) in the ventral nerve cord of the oligochaete Pheretima communissima (gf, giant fiber).** [Redrawn after F. Ogawa, *Sci. Rept. Tokohu Imp. Univ.*, *13* (1939), 395–488.]

The choice of location for ganglia seems to depend on the configuration of the body and the desirability of placing relay stations in regions where great and special activity takes place. As already mentioned, this is preeminently the front end, that end of the animal first exposed to the vagaries of the environment, where large numbers of sense organs must be tied into the nervous system. This portion of the central nervous system remains dorsal to the alimentary canal, while the rest of the system is ventral. The two are connected by nerves going around the gut. The greatest development occurs in the dorsal region (supraesophageal, or cerebral, ganglia) and the next greatest immediately below (subesophageal ganglia). The degree of development of anterior ganglia is closely correlated with the complexity and mass of sensory equipment. Thus the head ganglia of free-living roundworms are well developed compared with those of parasitic worms, the trematodes and cestodes, whose sense organs are few.

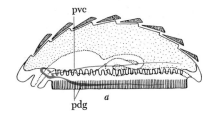

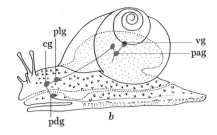

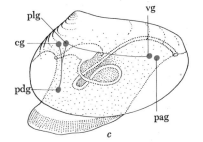

Figure 4.7 *Molluscan nervous systems. (a) A chiton. (b) A gastropod. (c) A lamellibranch:* cg, *cerebral commissure;* pvc, *pleurovisceral commissure;* pag, *parietal ganglion;* pdg, *pedal ganglion;* plg, *pleural ganglion;* vg, *visceral ganglion.* [Redrawn from L. H. Borradaile and F. A. Potts, *Invertebrata.* New York: Crowell Collier and Macmillan, Inc., 1935.]

Figure 4.8 *Anterior part of the nervous system of the earthworm (Lumbricus terrestris):* m, *mouth;* bc, *buccal cavity;* ce, *cerebral ganglion;* pm, *prostomium;* ph, *pharynx;* sg, *stomatogastric nervous system;* se, *subesophageal ganglion.* [Redrawn after W. N. Hess, *J. Morphol., 40* (1925), 235–261.]

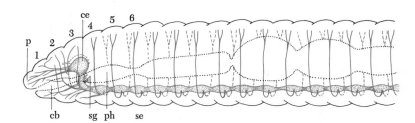

The subesophageal ganglion produces tone, spontaneity, search movements, and backward flight. The brain regulates these activities by directing them in response to special sensory input. It is responsible for variations in inner state, preference and choice of food, habitat, and so on. The ventral nerve cord regulates locomotion, righting, twitch response, and withdrawal. Where extensive motor activity is called for, ganglia tend to arise as motor relay stations. The pedal ganglion in the razor clam (*Ensis*) is an example. The anal and genital regions are also sites of ganglion formation (for example, the perianal ganglion of *Ascaris*). The placement of ganglia in strategic locations is most strikingly illustrated by the molluscs, which have at least three pairs: cerebral, pedal, and visceral (Figure 4.7). Animals that are markedly segmented (annelids and arthropods) initially developed segmentally arranged ganglia (Figure 4.8).

BILATERAL NERVOUS SYSTEMS

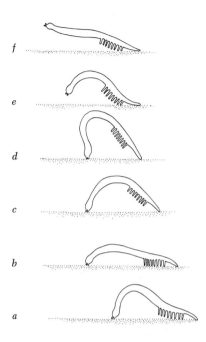

f

e

d

c

b

a

Figure 4.9 Inchwormlike locomotion of a free-living nematode (draconematid). [Redrawn from L. H. Hyman, *The Invertebrates.* New York: McGraw-Hill Book Company, 1951, Vol. 3. After H. Stauffer, 1924.]

The role of ganglia in relaying sensory activity to effectors and in ordering or coordinating the action of the effectors they serve is most clearly seen in the locomotor activity of worms. Flatworms, which lack a complicated musculature, are well served by a nerve net and ganglionated cords. Since the skeletal development of roundworms permitted a highly organized musculature, intricate modes of movement became possible. The draconematid worm, for example, is able to progress in an inchwormlike fashion because it is able to coordinate the number and sequence of muscles that contract (Figure 4.9). In a segmented animal, even more versatile locomotion is possible. How this is accomplished in the earthworm is illustrated by the ingenious experiment of Friedlander shown in Figure 4.10.

The earthworm moves by means of waves of peristalsis passing from anterior to posterior. If the entire ventral nerve cord is removed, locomotion is not possible. This proves that activity is not propagated by the subepidermal nerve net. If a worm is transected completely except for the nerve cord, which is the only part that holds the two ends of the worm together, a coordinated peristaltic wave still occurs. However, its persistence when the cord is cut or even removed from several segments shows that it need not be transmitted via the cord. If a completely transected worm is held together by threads (Figure 4.10), the wave is essentially normal. This clever experiment illustrates that as each segment contracts, it exerts traction on the succeeding segment (via the threads) and stimulates its sense organs (proprioceptors). A pattern of reciprocal excitation and inhibition is probably inherent in the cord but to execute locomotion requires maintained stimulation either from the substrate or from stretching of the body.

A different type of locomotion occurs in the sea worm *Nereis*, which moves in a wavy, snakelike fashion. Here the coordination of muscle contraction is such that the muscles on one side of a segment contract while the corresponding set on the other side relaxes; this action is alternated. Leeches move by an inchwormlike action that requires still more complex coordination; the anterior and posterior suckers alternately attach and disengage in coordination with the contractions of the body musculature. In these forms, an intact cord and the ganglion serving the motor area are necessary, but the cerebral ganglion is not required. Similarly, the cerebral ganglion is not required by the mussel (*Mytilus*) when it moves and spins threads for attachment to the substrate, using the visceral ganglion. Clearly, it is possible for each area of these animals to be more or less autonomous and still act in concert with other parts without a master control center. The potentialities of this arrangement are definitely limited.

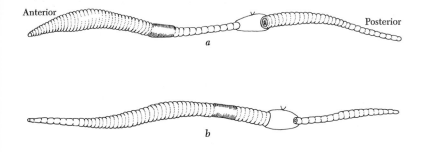

Anterior

Posterior

a

b

Figure 4.10 **The experiment of Friedlander (1888) demonstrated that a contraction wave would pass from the posterior end to the anterior end of a worm even though a gap of a centimeter separated two halves, provided that the halves were joined by a thread.** [Redrawn from P. P. Grassé, *Traité de Zoologie.* Paris: Masson, 1959, Vol. 5.]

Ganglia (and ganglionated cords) act not only as relay stations but also initiate activity. Thus, complicated adaptive movements may be carried on by a series of spontaneously active "clocks." The lugworm (*Arenicola marina*), for example, lives in a U-shaped burrow which it keeps open by rhythmic respiratory, locomotory, and feeding movements. These rhythmic cycles (the defecation cycle, too) are not simple reflexes in response to a changing environment; they are rhythmic muscular contractions driven by spontaneous activity in various parts of the nervous system. The brain is quite unnecessary; isolated fragments of the animal perform their normal cycle well.

As the nervous system condensed in the course of evolution from a diffuse net, it was no longer functionally similar throughout. The increasing importance of ganglia in controlling those areas of the body that they innervate finally demanded that they themselves come under some higher control; otherwise the animal would have to act like a republic of parts rather than as a unit, and chaos would prevail. The role of master control center fell to the anterior ganglia, especially the supraesophageal (cerebral) one.

Originally, cerebral ganglia were principally sensory relay centers (Figure 4.6). Oligochaete worms (earthworms), for example, can still eat and burrow in normal fashion after removal of the cerebral ganglia. The simplest type of control that cerebral ganglia exercised was that of excitation and inhibition of other ganglia. In polychaete worms, the head ganglia (Figure 4.11) are more than relay stations. The supraesophageal one is an inhibitory center; the subesophageal one, a motor center. Without the former, *Nereis* no longer feeds or burrows, is less sensitive to light and chemicals, and is hyperactive. Without the subesophageal ganglion, it is inactive (akinetic).

In gastropod molluscs (snails, limpets, and so on), the cerebral

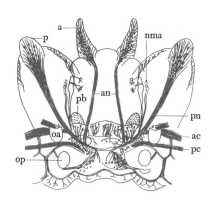

Figure 4.11 *Cerebral ganglia in a polychaete worm (Nereis diversicolor): a, antenna; an, antennal nerve; pc, ac, periesophageal commissures; p, palp; pn, palpal nerve; pb, portion of pedunculate body; oa, op, eyes.* [Redrawn after G. Retzius, *Biol. Untersuch., 7* (1895), 6–11.]

INTEGRATION AND
COMPARTMENTALIZATION

Figure 4.12 *Transverse section through the cerebral ganglion of the ascidian Phallusia nigra, showing the concentration of ganglion cell bodies in the cortical region.* [Redrawn from W. A. Hilton, *J. Entomol. Zool. Claremont, 33* (1941), 44–53.]

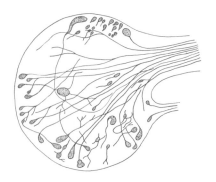

(supraesophageal) ganglion has clearly become the controlling center for the activities of the animal. As a consequence of its development, these animals are capable of behavior of considerable complexity. In addition to habituation and simple conditioning, they exhibit complicated coordinated reflex behavior and a marked ability to modify their behavior. The snail *Helix,* for instance, displays a complicated courtship behavior in which two individuals approach, evert their genital areas, and launch calcareous darts with enough force to penetrate each other's internal organs. Following this heroic stimulation, fertilization occurs.

Some snails learn a T maze after about 60 trials and retain the habit for about 30 days. Other gastropods, notably limpets, show a homing sense that suggests true learning of topographic relations. Limpets, which fasten themselves to a particular spot on a rock and make feeding excursions from there at low tide, usually return to the same spot. Although they may return along the outgoing path, they are not dependent on it and may return from a distance as great as 5 ft by an entirely different route. This behavior would tend to rule out the use of trail-following cues. It indicates that these animals can make a well-integrated assessment of sensory cues.

Ganglia began as relay stations. Later they started to modify the messages they received before passing them on to motor systems. They also began to initiate activity, which they passed on. They acquired these accomplishments when they ceased to be mere aggregations of sensory and motor cell bodies whose axons stretch out to the farthest reaches of the body. Indeed, many new cells were added whose processes never extend beyond the confines of the ganglion. Ganglion development was characterized by multiplicity and complexity. Not only was the number of neurons increased, but the kinds of neurons increased as well. An infinite number of varieties differing in size, shape, and number of processes, in shape and size of cell body, and in the number and nature of connections arose. Differences in structure reflect differences in function. And the cells of ganglia then gathered together into tight groups, which enhanced the number of possible connections in a given space and reduced the time of transmission.

As ganglia became compartmentalized, function became localized in different areas, ganglia became integrative centers of great complexity, and behavior extended to new horizons. One need only compare the simple ventral ganglion of an oligochaete worm (Figure 4.6), the elementary cerebral ganglia of a polyclad flatworm (Figure 4.3) and those of one of the chordate ancestors of the vertebrates (Figure 4.12), the more complex ganglion of a polychaete worm (Figure 4.11), of the insect (see Figure 5.6), and of the vertebrate (Figure 6.1) to appreciate the significance these developments must have had on behavior. And in our preoccupation with neurons, we must not overlook a significant new development, beginning in annelids: the incorporation into ganglia of cells (neurosecretory) whose secretions have profound effects on the behavior of animals.

Coincident with the development of complex means of locomotion has been the elaboration of mechanical senses to detect stress and strains within the body and to signal the postural relation of one part of the body to another. As a consequence, behavior consists not only of locomotion and feeding but also of characteristic postural attitudes (for example, the positions assumed by earthworms during copulation) and of manipulation of the environment.

Despite the lack of appendages (in earthworms) or anything more than mere fleshy flaps (parapodia of seaworms), a certain dexterity has been attained by worms. Darwin was fascinated by the apparent intelligence earthworms show in dragging leaves into their burrows. Emerging from its burrow at night, an earthworm keeps its posterior end firmly anchored in the gallery as it stretches its anterior end across the surface of the ground in exploratory movement. When certain vegetable debris is encountered, it is pulled over and into the burrow by the proboscis. Darwin believed that leaves (or bits of paper cut into various shapes) were always seized in such a way as to insure that the shape offered the least resistance to passage into the burrow. Although it is true that the worm can differentiate the stem from the other end of a leaf by chemical stimuli, recent studies have shown that the successful pulling of leaves into a burrow is a trial-and-error process. By and large the actions of earthworms are simple, rigid responses to stimuli.

Many aquatic worms construct quite serviceable tubes in which they live. One such (*Aulophorus carteri*) builds a tube with the spores of an aquatic plant (Figure 4.13). The anterior end of the worm undulates until the prostomium encounters a spore. When this happens, the head is flexed and the spore is taken into the mouth and covered with saliva. The worm then retracts to bring the spore to the top of the tube, where it is held in place until firmly stuck.

Some seaworms (for example, *Chaetopterus*) secrete mucus that is manipulated to form a bag which serves as a strainer for removing food from water circulated through it. When the bag is full, it is rolled up, transferred to the mouth, and swallowed.

Figure 4.13 *Method of tube construction by the worm Aulophorus carteri, using the spores of an aquatic plant.* [Redrawn from P. P. Grassé, *Traité de Zoologie.* Paris: Masson, 1959, Vol. 5. After Carter and Beadle, 1931.]

In addition to the various kinds of activity described, worms exhibit twitch reflexes, reflex arrest of creeping, opening and closing of the anus and various pores, and defensive proboscis protrusion. Marine polychaetes especially indulge in vigorous aggressive behavior in which two animals, upon meeting, thrust forward the proboscis very rapidly, seize one another with the jaws, and jerk and twist the body much as a dog worries a bone. Both intraspecific and interspecific aggression occurs.

The level of neural organization characteristic of worms permits not only habituation but fairly facile conditioning, especially to light, shadow, touch, and electric shock. For this level of learning the brain is not needed. Nor is it required for retention. Decerebrate worms can learn T mazes and be conditioned (see page 118). The brain is not totally irrelevant, however, because conditioning can occur in its absence only if the intertrial intervals are shortened. Nevertheless, it is clear that learning can occur in other parts of the nervous system. The responses are withdrawal actions or approach to food or chemicals. The general form of the conditioning and the variables that affect it are similar to what is observed in other animals, vertebrates included.

RELATION BETWEEN BEHAVIOR AND THE
ACTIVITY OF GANGLION CELLS

The sea hare *Aplysia* is a gastropod that, like other rather simple molluscs, shows rhythms, habituation, and classical conditioning. Electrophysiological analyses of its ganglion cells are beginning to reveal some of the neural mechanisms underlying these kinds of behavior. There are, for example, single cells in the abdominal ganglia that show circadian rhythms of spontaneous firing that are related to the animal's prior light-dark regime. A cell of this type retains its rhythm of firing after the ganglion is removed from the animal. If the light-dark regime of an animal is shifted, a ganglion cell reflects the shift in an altered rhythm. The rhythm has a 2-week period whose oscillation lags slightly behind the time of high tide at the locality were the animal was captured. This example of a biological clock at the neuronal level is clearly a genetically based behavior program that can be modified by experience and can store a record of this experience.

Habituation can also be detected at the neuronal level. When drops of water fall on the head of a sea hare, it contracts. After repeated drops at 30-sec intervals, the animal no longer contracts. Dishabituation occurs after a rest of 10 min. It can also be brought about by scratching the sea hare on another portion of its body. In an isolated preparation consisting of the head and an upper ganglion with a microelectrode to record postsynaptic potentials in one cell, a drop of water on the head

Figure 4.14   Type 1 illustrates the relationship between stimulus sequences in two neural pathways and a classical conditioning paradigm. Type 2 represents the situation for instrumental conditioning. The middle sections (analogs) illustrate how the conventional statements on conditioning (top section) can be applied to isolated ganglia when electrical stimuli are employed in place of behavioral ones. In Type 2 the reinforcement can come when the cell begins its spontaneous burst (contingency A), in which case the next burst is hastened, or when the cell is quiet (contingency B), in which case the cell delays its next burst. The cell is thus "trained" to give either frequent or infrequent bursts of response. The modifications produced by Type 1 conditioning are shown in Figure 4.15, where pathway 1 and pathway 2 are stimulated in succession. [From E. Kandel, N.R.P. Bull., 4 (1966), 142.]

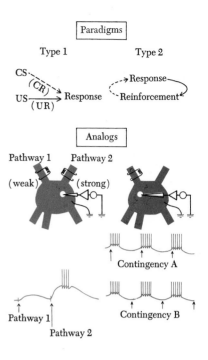

causes the potential to increase. After 15 drops the increase is very small. Following a rest of 15 min it returns to normal. Scratching the head also produces complete recovery. Here habituation and dishabituation are seen in a single ganglion cell when it is activated through a synapse.

Even more interesting is the demonstration of situations resembling classical and operant conditioning in single ganglion cells and their synapses (Figure 4.14.) In the parietovisceral ganglion, which contains 1,000 of the approximately 10,000 cells in the entire nervous system, certain individual cells can be identified. It is possible for two excitatory nerves to a single cell to be stimulated individually (Figure 4.15). One pathway produces only a small synaptic potential; this is analogous to the conditioned stimulus. The other produces both a larger potential and action potentials, and is analogous to the unconditioned stimulus. If the two stimuli are paired in the order CS–US for about 33 trials, an increase in the synaptic potential is then observed when the first nerve (CS) is stimulated alone. After 60 pairings, stimulation of the first nerve alone produces action potentials. This enhanced response only occurs if the stimuli are paired in the trials. The situation clearly is analogous to classical conditioning (page 110).

A situation analogous to instrumental conditioning has also been observed (Figure 4.14, Type 2). Here a cell that had spontaneous activity in the form of periodic bursts of action potentials was stimulated electrically every time a burst occurred. This reinforcement caused the next burst to occur earlier. Stimulation is thus analogous to reward in operant conditioning.

Instrumental conditioning at the behavioral level has been demonstrated in the isolated ganglion of cockroaches and locusts. An intact locust, for example, can be trained to flex its leg in order to avoid a mild shock. Records of electrical activity from the coxal adductor, the muscle involved in leg raising, revealed that small spontaneous dis-

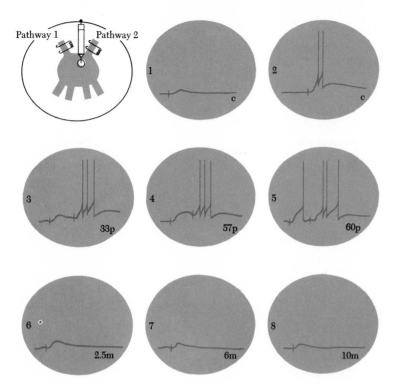

Figure 4.15 *The results of pairing a weak shock via pathway 1 with a strong shock via pathway 2. Number 1 shows normal response to weak shock and number 2 shows normal response to strong shock before pairing. The postsynaptic potential is small in 1 and large in 2. Spikes are generated. Numbers 3 to 5 show results of 33, 57, and 60 pairing trials, respectively. Note build-up of postsynaptic potential to the weak shock and the eventual generation of an action potential (5). Numbers 6 to 8 illustrate decline of postsynaptic potential after pairing ceases.* [Redrawn from E. Kandel, *N.R.P. Bull., 4* (1966), 143.]

Figure 4.16 *Isolated preparation from locust consisting solely of a ganglion and a leg muscle. Symbols E and 1-C represent, respectively, the single excitatory and inhibitory axons that innervate the muscle. The positions of various stimulating electrodes are indicated by parallel dark bars. Stimulation can be applied electrically via these or mechanically by tweaking with forceps. The intracellular electrode records spontaneous activity in the muscle.* [Redrawn from G. Hoyle, *J. Exp. Biol., 44* (1966), 413–427.]

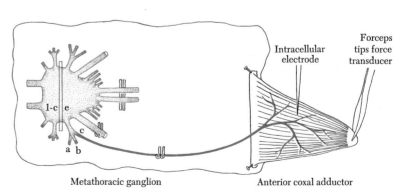

Metathoracic ganglion

Anterior coxal adductor

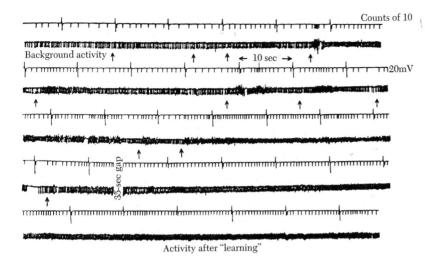

Counts of 10

Background activity

← 10 sec →

20mV

35-sec gap

Activity after "learning"

Figure 4.17 **"Learning" in the isolated locust preparation.** *Shocks (arrows) were applied to the leg whenever the experimenter judged that the rising frequency of spontaneous activity in the leg muscle began to fall. At the start, the background activity was 6 spikes per second. After 11 timed shocks were applied over a period of less than 4 min, the spontaneous activity rose to 20 spikes per second.* [Redrawn from G. Hoyle, *J. Exp. Biol., 44* (1966), 413–427.]

charges that occurred before training were greatly increased after learning. An isolated preparation consisting solely of the metathoracic ganglion, the crural nerve, and the coxal adductor responded to tweaking (by forceps) of the muscle with the same increase in electrical activity that was shown by the intact preparation subjected to shock (Figure 4.16). When the small spontaneous discharge was carefully monitored and a shock applied to the nerve every time the frequency began to wane, an increase was produced that could be maintained for considerable periods of time (Figure 4.17).

Cephalopods have the largest brains of all invertebrates. The supraesophageal ganglia have burgeoned forth and, in association with the subesophageal ganglia, form a complex and highly talented brain composed of, in the case of the octopus, about 168 million cells. More important than size, however, is the great elaboration of texture and differentiation of the neuropile, almost equal to that of the highest arthropods and fishes. This is a complex synaptic field, and regional differentiation is extensive. As a consequence of this development, cephalopods are capable of behavior that far outstrips that of any of their evolutionary predecessors. They exhibit complicated postural and fright behavior, sexual display (in *Sepia*), and intricate copulation behavior. The octopus, for example, inserts one arm into the mantle cavity of the female and deposits a spermatophore there. The octopus con-

CEPHALOPODS (SQUIDS, CUTTLEFISH, OCTOPUSES)

BILATERAL NERVOUS SYSTEMS

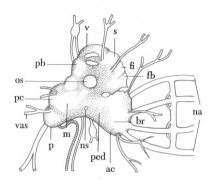

Figure 4.18 *The brain of Octopus, dissected out and seen from the right side, with the optic stalk cut and the optic lobe removed: os, optic stalk; pb, posterior basal; pc, posterior chromatophore; vas, vasomotor; m, magnocellular; p, pallioviscera; ped, pedal; ns, nerves to the statocyst; ac, anterior chromatophore; br, brachial; na, nerves to the arms; fb, frontal buccal; fi, frontal interior; s, superior; v, vertical.* [Redrawn from M. J. Wells, *Brain and Behaviour in Cephalopods.* Stanford, Calif.: Stanford University Press, 1962, p. 93.]

structs a home from debris on the ocean floor; it has territorial behavior, that is, it confines its activities to a circumscribed area and behaves aggressively toward other individuals entering this area; it has visual discrimination and form perception equal to or exceeding that of insects; it can be conditioned readily and learns a maze with ease. Its learning abilities greatly transcend those of any animal discussed thus far.

All these advances in behavior, and the limits as well, are indisputably associated with the development of the brain. The contributions of different parts of the brain to behavior have been studied by electrical stimulation and by making lesions. They clearly demonstrate great subdivision of function. Complexity, multiplicity, and compartmentalization have been carried to the point where there are fourteen main lobes mediating different functions (Figure 4.18). The anatomically lower lobes regulate only simple functions; the sensory lobes (for example, optic lobes) receive, discriminate, and analyze stimuli from the environment and appropriately activate the motor centers; the highest centers (anatomically and functionally) receive activity from sensory ones, mediate complex behavior, and monitor and regulate the entire system.

We can investigate the contributions of the various centers to behavior by making cuts and lesions in various parts of the brain. An octopus with only its subesophageal lobes (Figure 4.19a) is like a spinal

Figure 4.19 *Diagram illustrating the effect of removal of various parts of the brain on the posture and movements of the octopus. (a) Complete removal of supraesophageal lobes. (b) Complete removal of supraesophageal lobes, leaving opticosubesophageal connections intact. (c) Severance of optic tracts and removal of half of the supraesophageal ganglia. Symptoms following bilateral optic tract section are due to postoperative shock.* [After B. B. Boycott and J. Z. Young, "Physiological Mechanisms in Animal Behaviour," in *Symposia Soc. Exp. Biol., 4.* New York: Academic Press, Inc., 1950.]

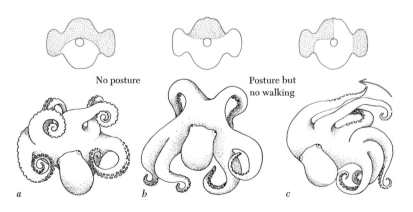

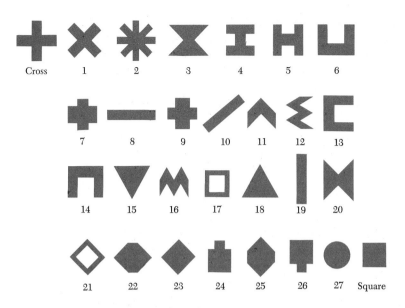

Cross	1	2	3	4	5	6

7	8	9	10	11	12	13

14	15	16	17	18	19	20

21	22	23	24	25	26	27	Square

Figure 4.20 **Octopuses were trained to discriminate between the cross and the square. Intermediate shapes were shown to the animals, and the number of attacks were recorded. The shapes are arranged in order according to the number of attacks made on each, for example, 1 and 2 were attacked most by animals trained to a cross, and 26 and 27 were attacked most by animals trained to a square.** [Redrawn from N. S. Sutherland, *J. Comp. Physiol. Psych.*, 55 (1962), 939–943.]

vertebrate (one in which only the spinal cord is functional) in that it is capable only of simple reflex movements. If these lower motor centers retain their connections with sensory lobes (for example, optic ones) but are isolated from the rest of the brain (Figure 4.19b), the animal maintains a rigid posture much like that of a decerebrate vertebrate (one with part of the brain removed). With the lower centers plus the optic lobes plus one-half of the supraesophageal lobes (Figure 4.19c), the animal can walk, but only in circles.

Compartmentalization of function is carried to a fine degree. Within the supraesophageal area of the brain the basal lobes are higher motor centers that may be likened to the midbrain of vertebrates. They initiate complex movements of the head and arms, inspiration, and so on. The motor control system is organized into a hierarchy in which each level is responsible for responses of differing degrees of complexity. Suckers on an isolated arm respond reflexly almost as normally as when the arm is attached. An isolated arm performs about as well as an attached one; it carries out all of the coordinated movements necessary for grasping food and passing it to the mouth (in this instance to the place where the mouth would normally be). Fragments of animal containing still more of the nervous system, as, for example, the subesophageal parts, are capable of integrated behavior in which all of the arms respond together. More complicated responses (walking and swimming) are possible if all of the brain except the optic lobes is present. Removal of those parts of the brain concerned with learning does not affect the organization of motor responses.

The five centers lying above the basal lobes are concerned with still more complex nuances of behavior and with learning. One of these (the vertical) is analogous in many respects with the cerebral cortex of mammals. Octopuses can learn many kinds of visual and tactile discriminations with an accuracy comparable to that exhibited by mammals. An octopus in captivity normally rests quietly until it sees a small moving object. It then attacks. By means of rewards (bits of fish or crabs) and punishment (electric shock) it can be trained to execute a wide variety of visual discriminations. For example, an octopus was first trained to approach and seize a crab lowered into the water by a thread. Then on one-half of the trials a white card was lowered with the crab, and in these trials the octopus was driven away by strong electric shock. After 12 trials the octopus begain to hesitate in its approach and by 24 trials it consistently remained in its nest when the card was present and consistently came out to feed when the crab was presented alone.

Visual learning depends on the optic, superior frontal, and vertical

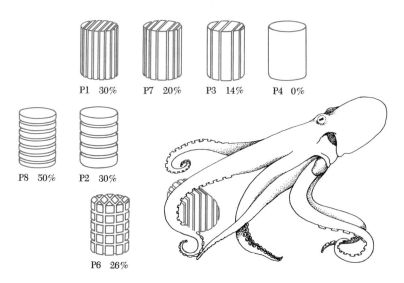

Figure 4.21 **Octopuses can be trained to discriminate tactually among members of top row and between P2 and P8. Forms P1, P2, and P6, having almost the same proportion of surface cut away, are apparently alike to them.** [M. J. Wells and J. Wells, *J. Exp. Biol.*, 34 (1957), 131–142.]

P1 30%   P7 20%   P3 14%   P4 0%

P8 50%   P2 30%

P6 26%

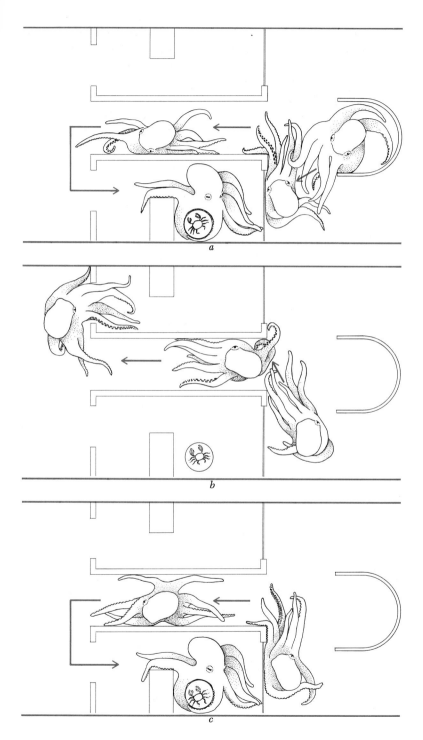

Figure 4.22 **Apparatus for testing ability of octopuses to solve detour problems.** (a) Octopus leaves home at right, struggles for 17 sec against window of feeding area, then passes down opaque corridor and around into feeding compartment. (b) Error by octopus blinded in right eye. It found the "wrong" wall and was led astray. (c) A correct run by the animal blinded in the right eye. Numbers indicate successive positions: 2 is 1 sec after 1; 3 is 3 sec after 2; 4 is 13 sec after 3. Positions 5–13 at 2 sec intervals. [Redrawn from M. J. Wells, *J. Exp. Biol.*, 41 (1964), 621–642.]

lobes; tactile learning on the superior frontal-subfrontal-vertical lobe system. The vertical lobe is in part a memory system (see page 124). When it is removed, animals that have been trained to do a task have to be retrained. This lobe is also presumed to distribute to all parts of the optic lobes the representations of visual stimulus patterns being learned.

Octopuses can discriminate among many geometrical shapes. For example (Figure 4.20), animals trained to attack a cross attacked shapes 1 and 2 most often and successive shapes fewer times; animals trained to attack a square attacked shapes 27 and 26 most often. In this discrimination the area-to-outline ratio seems to be the critical dimension. There is no doubt that the visual acuity and capacity for being able to recognize small differences in shape is related to the complexity and arrangement of neurons in the optic lobes.

The sense of touch is also an important avenue of information. Blinded octopuses can be trained in as few as 20 trials to detect differences in texture. Differences in size and shape are detectable only if they cause different degrees of distortion of the suckers on the arm in contact. A large object with an irregular surface cannot be distinguished from a small smooth one. Differences in weight, surface patterns (Figure 4.21), and shape cannot be discriminated. Part of this failure is probably attributable to octopuses' inability to utilize information about their posture or the relative positions of parts of their bodies in learning. Sensory input of this sort apparently is not relayed to those parts of the brain involved with learning. This conclusion is consistent with what lesion experiments tell about the organization of the motor system. It would deny the animal the ability to learn skilled tasks as, for example, unscrewing a bottle top.

Recent experiments now reveal that octopuses are able to master "detour" problems. Figure 4.22 (a, b, and c) illustrates an experimental situation. The octopus in its home can see a crab through a glass partition. To get the crab it must go down an opaque central corridor and then make two right or two left turns. The octopus not only remembers that there is a crab but where it is. The results of two experiments indicate that the octopus probably does not solve the problem by remembering the bodily movements it has made. First, removal of the statocysts (the principal rotation receptors) upsets equilibrium but does not impair the ability to solve "detour" problems. Second, animals blinded on one side veer 180 degrees off course and cannot learn to correct the mistake: the octopus follows whatever wall it happens to come into contact with until it locates breaks. This is further evidence that the animals are not able to accurately appreciate the orientation of their bodies in space.

WITH THE EXCEPTION OF CEPHALOPODS, THE ARTHROPODS EXHIBIT A
richer repertory of behavior than do any other invertebrates. Three
developments have made such complicated behavior possible: the

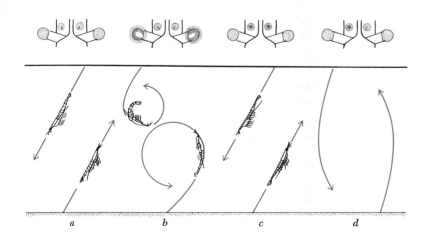

*Figure 5.1 **The importance of eyes
and statoliths for normal swimming
by crustacea (Palaemonetes and Cran-
gon): (a) eyes normal, light from
above, no statoliths, upward paths
spontaneous or induced by CO_2; (b)
eyes darkened except dorsally, light
uniform and diffuse, no statoliths; (c)
statoliths normal, light diffuse; (d) one
statocyst lacking its statolith, light diffuse.** [Redrawn from H. Schöne, The
Physiology of Crustacea. New York:
Academic Press, Inc., 1961, Vol. II.]*

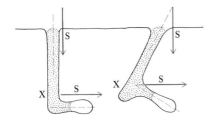

Figure 5.2 *Burrows made in a container by two fiddler crabs (Uca pugilator). When the crabs had dug to point X, the container was rotated 90 degrees. The symbol S indicates the direction of gravitational force.* [Redrawn from H. Schöne, *The Physiology of Crustacea.* New York: Academic Press, Inc., 1961, Vol. II. After J. B. Demboroski, *Biol. Bull., 50* (1926), 179–201.]

elaboration of very complex sense organs, which permit a highly discriminative assessment of the environment; the evolution of jointed appendages and their subsequent modification into legs and mouth-parts of extraordinary complexity, making possible exceptional manipulative ability; the development of a brain that is complex enough and has a sufficiently integrative capacity to organize the wealth of sensory information received and to direct all the motions of appendages. Compound eyes and statocysts seem to be the dominant sense organs in crustacea. The chemical senses are not so well understood in these animals as they are in insects so the extent of the contribution to behavior cannot yet be fully assessed. The importance of vision is obvious when one notes the kinds of behavior executed (See Figures 5.4 and 5.5). It is strikingly demonstrated by the observation that a fiddler crab will fight its own image in a mirror. The importance of statocysts, especially in matters of equilibrium and directional orientation (for example, swimming, digging of burrows) is easily demonstrated experimentally (Figures 5.1 and 5.2).

The central nervous system of crustaceans is characterized by a small number of cells (97,722 in the crayfish) and elaborate interneurons that fulfill the function of whole tracts in vertebrates. Many aspects of behavior, but by no means all, can be explained in terms of reflexes—for example, eyestalk-withdrawal, claw closing and opening, escape, defense, feeding, copulation (the last two can be shut off by the brain), and righting. There is, however, much more complex behavior that still awaits analysis.

The great complexity and diversity of crustacean behavior is revealed in the varied feeding habits, grooming patterns, nest building, camouflage techniques, and sexual, parental, and social activities. Crustacea may be carnivorous, vegetarian, or omnivorous. Predatory species ambush, stalk, and chase prey. Some simply wait in the doorways of their burrows and strike any small creature that passes within reach while others actually pursue swimming shrimp and small fish. Many predatory crabs are quick and agile enough to capture flies. One omnivorous species (*Birgus latro*) climbs pandanus trees for the fruit, which it may carry to a hiding place hundreds of meters away.

Most crustaceans carefully and thoroughly groom themselves although the amount of time spent in this pastime varies from species to species. In some crabs the appendages that are employed in grooming are equipped with special brushlike margins. The fiddler crab *Uca* is often seen cleaning mud from its large claw by means of the smaller one. After a molt, at which time the statoliths are lost (see Figure 2.7) many crabs pick up small grains of sand with their cleaning claws and stuff them into the statocysts.

Although many crustaceans seek protection wherever they happen to be by scurrying under a stone or digging themselves into the ocean bottom, others construct more or less permanent shelters. Lobsters and crayfish, for example, maneuver stones around to provide a "home" for themselves. Many species dig burrows, occasionally with lateral shafts for storage. The burrow is terminated at a depth where ground water is encountered. The opening of the burrow may be closed by a plug either constructed outside and lowered into place from within or made by molding the side of the entrance together and reinforcing with mud dug from the bottom (Figure 5.3). Another crab "spins" a dome-shaped shelter by rotating rapidly, all the while pushing pellets of sand around and above itself until a sand igloo is completed. The direction of digging is regulated by gravity. If a crab is digging in a container of sand and the container is rotated 90 degrees, the direction of the shaft changes by 90 degrees. Removal of the statocysts also interferes with the orientation of digging (Figure 5.2).

Many crabs go through astounding acrobatics to camouflage themselves with other organisms. They may attach fragments of hydroid or bryozoan colonies, bits of sponges, filaments of algae, and entire sea anemones (Figure 5.4). Others rely on defense to protect themselves. Some run, then stop and fence with their claws; one species holds sea anemones in its claws and thrusts them toward the attacker. Some crabs seize an attacker in their claws, then autotomize the claw leaving it firmly attached to the predator while they run away.

Interactions between individuals and among members of groups are frequently very complex. Courtship may be elaborate. Parental care is very advanced in some species. Many levels of social behavior have evolved among crabs. Fiddler crabs, for example, have organized communal feeding groups, familial groups with dominant males, peck orders, and ceremonial and fixed patterns of fighting among males. As might

 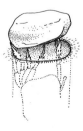 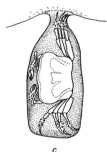

a   b   c

Figure 5.3 **Construction and closure of burrow by the fiddler crab (Uca).** [Redrawn from H. Schöne, *The Physiology of Crustacea.* New York: Academic Press, Inc., 1961, Vol. II.]

Figure 5.4   *Acquisition of an anemone by Dardanus arrosor.* [After F. Brock, *Arch. Entwicklungsmech. Organ., 112* (1927), 204–238.]

Figure 5.5   *Signaling gestures of two species of fiddler crabs (Uca).* [From J. Crane, *Zoologica, 42* (1957), 69–82. Courtesy of N. Y. Zoological Society.]

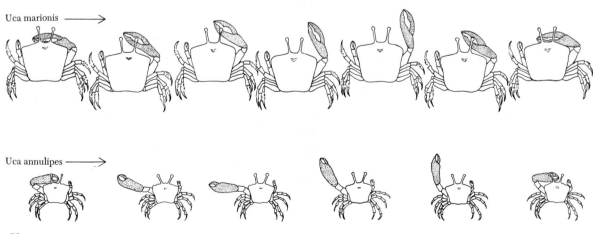

Uca marionis

Uca annulipes

be expected there are efficient means of communication. Stridulation is employed as a warning sign. Tactile stimuli are also used. Visual signaling is probably the most important means of communication and each species has its characteristic gestures (Figure 5.5). There is no evidence that the high level of chemical communication attained by insects has evolved in crustacea. Chemical communication is probably less well suited to an aquatic environment than to a terrestrial one.

The apogee of invertebrate behavior is realized in the insects, especially in the Hymenoptera (ants, bees, and wasps). There are over a million species in the class, and one would hardly expect either the kind or the level of behavior to be uniform in the group. Consequently, we shall examine only the highest behavioral achievements, which illustrate what can be done with the kind of nervous equipment available, and we shall also analyze the relationship between behavior and the neural developments we have been discussing.

It is clear from the mountain of literature on the subject that extraordinarily complex behavior is possible. Much of this behavior resembles and even rivals that of mammals, so much so that writers have been led to impute to insects reasoning and intelligence. This is the greatest tribute to the complexity of their behavior; however, experimental analyses show that these are largely stimulus-bound animals that operate in a stereotyped fashion in strict accordance with the stimuli received At the same time, it must be emphasized that in the higher forms there is some plasticity of behavior and learning attains an important level.

The same evolutionary development that advanced the behavioral capacities of crustacea makes possible complex behavior in insects. In addition, the brain has become even more complex (Figure 5.6). The visual, olfactory, and tactile senses have become especially well developed. The compound eye, the best visual system developed by the invertebrates (with the possible exception of the cephalopod eye), may be deficient in visual acuity, form perception, and color vision by vertebrate standards, but it still possesses these qualities to some extent. It is superior in the perception of flicker, an attribute of considerable adaptive value because of the speed with which many insects fly. The olfactory sense, at least insofar as acuity is concerned, rivals the best that vertebrates have to offer. The mechanical senses are well developed, the proprioceptive senses necessarily so in order to keep the nervous system informed about the positions of the many parts of the highly articulated body; the tactile sense is developed to a degree that makes a highly accurate sensory survey of the substrate possible.

Figure 5.6 **Simplified diagram of the main areas of the insect brain and their principal fiber tracts:** *on,* optic nerve; *ol,* optic lobe; *ot,* optic tract; *cc,* corpus centrale; *cp,* corpus pedunculatum or mushroom body; *pc,* pons cerebralis; *pi,* pars intercerebralis; *pcl,* protocerebral lobe; *al,* accessory lobe; *an,* antennal nerve; *ac,* antennal center; *1 com, 2 com, 3 com,* commissures; *fg,* frontal ganglion; *csg,* circumesophageal connectives to subseophageal ganglion. [Redrawn from R. E. Snodgrass, *Principles of Insect Morphology.* New York: McGraw-Hill Book Company, 1935.]

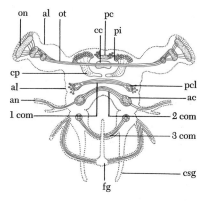

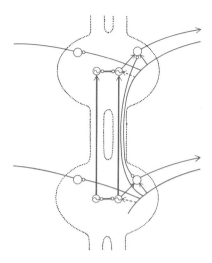

Figure 5.7   A scheme (not an actual neuronal network) to illustrate how ganglionic oscillators and reflexes might operate to regulate walking. Excitatory pathways (arrows) and inhibitory pathways (•) are shown between two ganglia and associated legs. Each half-ganglion has an oscillator (∼); contralateral ones are reciprocally connected. [Redrawn from D. M. Wilson, in Invertebrate Nervous Systems, C. A. G. Wiersma, ed. Chicago: University of Chicago Press, 1967, 219–229.]

Equipped with these systems, insects are able to manipulate their environment and to maintain elaborate relations with one another. They build nests of great complexity; burrowing bees and wasps hollow out underground nests and provision them with paralyzed prey as food for the young; mud-daubing wasps construct characteristic nests of mud which are also provisioned with prey; paper-building wasps erect elaborate nests from reconstituted wood fiber; honeybees construct precise waxen combs. In rare cases, insects use a bit of the environment as a tool. The classical case is that of certain tropical ants that make nests of leaves held together by silken threads. The ants themselves cannot spin, but the larvae can. The adults hold the larvae in their jaws and use them as shuttles.

Coincident with nest building is intimate association with others of the species. These associations culminate in highly organized social units, a strong sense of territoriality, and the acquisition of considerable homing abilities. Social insects—ants, for example—have morphologically distinct castes, each of which has its own characteristic behavior. This fact is especially interesting since all may be of the same genotype. The activities of the specialized castes may be quite bizarre, ranging from hunting, defense of the colony, and capturing and enslavement of other species, to cultivation of fungus gardens, care and tending of aphids (whose saccharine secretions are highly prized), and conversion of themselves into veritable wine casks for the storage of honey for the needs of the colony.

Territoriality involves the defense of a given piece of ground, the recognition of home, and the ability to orient to it from a distance. Insects' powers of orientation and their abilities to learn and remember landmarks rival those of the mammals. Walking forms (ants) deposit and follow chemical trails or orient by visual cues and the polarized light pattern of daylight. Flying forms (bees, wasps) orient by visual landmarks and can navigate over distances as great as a mile by determining the position of the sun and the plane of polarization of the light in the sky. Experiments in which landmarks have been moved proved that new landmarks can be learned very quickly, in one 9-sec orientation flight. Honeybees carry this ability further; in the hive they perform dances whose speed and orientation with respect to gravity code the distance and direction from the hive to the site of a given supply of food.

There are a number of cases in which insect behavior has been elegantly analyzed. Because these analyses have greatly advanced our knowledge of the relationship between behavior and underlying neural mechanisms, they will now be described briefly in the remaining sections of this chapter.

Insects may employ any or all of three basic modes of locomotion: swimming, walking, or flying. Of the many naturally occurring walking gaits, most can be depicted by a model which assumes the following: on each side the limbs move in the sequence—hind, middle, front; the two sides are out of phase; the interval between the stepping time of a front leg and the ipsilateral hind leg is variable but other intervals are more or less constant. Amputees adopt, without learning, a different gait, but this is not actually a *new* gait; it is a normal, slow-speed gait lacking simply the stepping of the missing legs. Coordination is not achieved simply by reflexes. There is evidence that endogenous ganglionic oscillators act together with reflexes. The simplest scheme supposes a pair of oscillators in each segment, reciprocally coupled and working antagonistically; they must also be intersegmentally coupled. Each oscillator drives two sets of antagonistic motor elements (Figure 5.7, with only one set shown). Proprioceptive input influences motor elements connected with the appropriate leg but does not affect the oscillator.

Flying also depends on ganglionic autorhythmicity. The complex sequence of muscular contractions responsible for the pattern of wing movements in locusts is programmed by the central nervous system and is markedly independent of external sensory or proprioceptive input. Steady wind blowing on special hairs on the head initiates flying in preparations totally lacking sensory input to the thorax. The flying pattern is normal, but the frequency is halved. To maintain maximum frequency, input from four stretch receptors at the base of each wing is required. This input enables the CNS to monitor the mechanical oscillation of the moving wings.

Sensory input does, however, alter behavior patterns of locomotion. Evasive behavior is a case in point (Figures 5.8 to 5.10). Some of the more important items of food for insectivorous bats are moths of the owl (noctuid), wooly bear (arctiid), and looper (geometrid) families. Bats capture these moths by echolocation, that is, they emit ultrasonic pulses that bounce off the flying moths, and they zero in on target by analyzing characteristics of the echo. The bats, however, do not score 100 percent of the time, and some of their errors occur as a result of moths' taking evasive action.

On each side of the thorax a moth has a tympanic organ consisting of two bipolar acoustic neurons. Both respond to sound in the same

Figure 5.8 **Tracks made by three moths in response to a train of ultrasonic pulses from a loudspeaker. The moth on the left clearly turned and flew away from the sound source, the one on the right dived while looping under power, and the one in the center shows a somewhat erratic response, even though it eventually turns away. The gaps on the tracks mark 0.25-sec intervals; the white dot near the beginning of each track marks the onset of the pulse train.** [Courtesy of Dr. K. D. Roeder.]

frequency range as that of bat cries by firing trains of impulses (Figure 5.10). One cell responds to low-intensity ultrasonics, hence can detect a hunting bat at distances as great as 130 ft; the other responds to higher intensity sound and detects bats within a range of 10 ft. In response to low-intensity sound, that is, to a distant foraging bat, moths make directional turns away from the source of sound (Figure 5.8). In response to high-intensity sound, that is, to a bat close at hand, moths make nondirectional, erratic, looping and diving flights (Figure 5.9). The astounding feature of the system is that only one bilateral pair of sense cells is necessary to initiate and steer the evasive behavior. With information from these two cells, vertical and horizontal localization of a sound source is possible.

At least six types of interneurons process information generated by the acoustic cells. One, the repeater neuron, distributes the signal faithfully to the contralateral side and up toward the brain. There are pulse-marker neurons (Figure 5.11) that fire only once per train (equals one sound pulse) of acoustic cell impulses and reset after a silent interval of 4 to 5 msec. The pulse-marker filters out sustained sounds, that is, it follows pulses above 20 per sec in a 1-to-1 fashion. There are still other neurons that sum input from right and left tympanic organs. There are train-marker units that fire steadily as long as an acoustic cell is firing and thus mark the duration of a train of sound pulses.

a

b

Figure 5.9   A tethered moth (Feltia subgothica) attempting to turn away from a sound. In a it is not flying; in b it is flying steadily in a forward direction; in c faint sound pulses came from the left side (S); in d the same signal came from the right side. Each frame was a multiflash exposure (about 8 flashes), hence the moth seems to have supernumerary wings and antennae. Turning away involves both bending of the antennae and body in the turning directions and partial folding of the wings on the off-side and extra extension of the wings on the near-side to the sound source. [Courtesy of Dr. K. D. Roeder.]

c

d

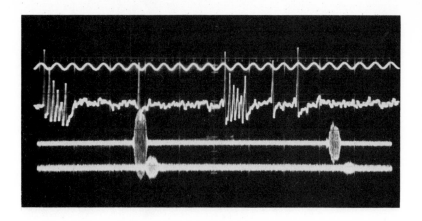

Figure 5.10 *Sample record of the pass made by a bat. Two of its cries are registered by microphones (bottom two traces). The response of the "ear" of the moth Catocala concumbens is shown in the second trace from the top. The top trace is 100 cps.* [Courtesy of Dr. K. D. Roeder.]

All of this neuronal integration serves to alter the rhythm and other characteristics of flight. Experiments on isolated ganglia of locusts have shown that the very complicated patterns of motor activity required to produce flying depend on interaction intrinsic to the collected motor neurons themselves. No particular pacemaker is required. Thus, as noted before, the basic command and organization of flight is resident in the ganglion.

Figure 5.11 *Record showing response of an acoustic cell and a pulse-marker neuron from the second thoracic ganglion of the moth Caenurgina erechtea to a sound pulse (lower trace) 35 msec in duration. The sound intensity was about +55 dB (0.0002 dyne/cm². The latency of the first A spike is about 4 msec; the pulse-marker appears at about 9 msec. Time is 5 msec per major vertical grid line.* [Courtesy of Dr. K. D. Roeder.]

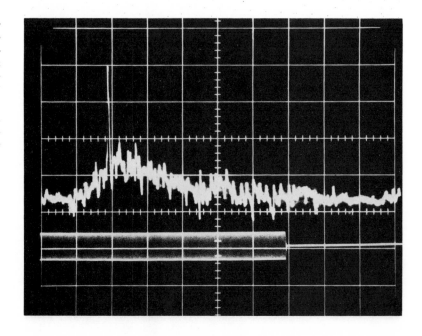

Unlike bats, mantids lie in ambush and wait for their victims to come to them. When a fly, for example, moves into the mantis's visual field, the mantis faces it by turning its head since its eyes are immovable. Usually the body is then turned to bring the head and prothorax in line although this alignment is not absolutely necessary. When the prey comes within proper distance either by its own movement or by careful stalking on the part of the mantis, it is captured by a lightning stroke (10 to 30 msec) of the forelegs. Since the stroke is so rapid, its direction must be determined in advance by information concerning the direction of the prey relative to the prothorax. There is not sufficient time for the mantis to control the stroke by watching the difference between its direction and that of the prey. This is the same problem that confronts a tennis player and may be taken as an example par excellence of absolute optical localization.

The mantis requires information about the prey relative to the head and the head relative to the prothorax. The first is provided by the eyes; the second, by proprioceptors in the neck (Figure 5.12). Normal mantids hit about 85 percent of the flies at which they strike (Figure 5.13). If information from the proprioceptors of the neck is eliminated by cutting the sensory nerve, hitting accuracy drops to 20 to 30 percent. If the head is immovably cemented to the thorax in the median position, performance remains normal. If the head is turned to the right and fixed, the prey is missed to the left, and vice versa. If the proprioceptors

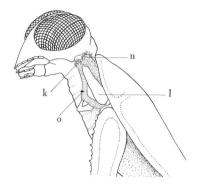

Figure 5.12 *Proprioceptors of the neck region of the mantis:* k, *sternocervical hair plate on anterior end of laterocervical sclerite* (l); n, *tergocervical plate. The common afferent nerve of both can be cut at* o. [From H. Mittelstaedt, in *Recent Advances in Invertebrate Physiology*. Eugene, Oregon: University of Oregon Press, 1957, p. 53.]

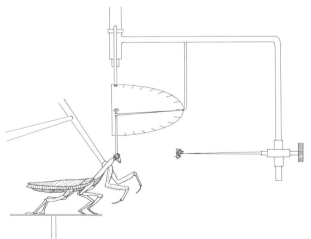

Figure 5.13 *Device for studying strike performance by the mantis. The animal is fixed at the prothorax, but the head is free to move about the vertical axis. Deviation of the head and of the prey from the median plane of the prothorax are measured on the same dial.* [From H. Mittelstaedt, in *Recent Advances in Invertebrate Physiology*. Eugene, Oregon: University of Oregon Press, 1957, p. 63.]

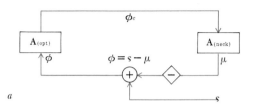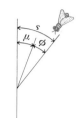

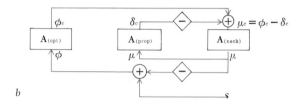

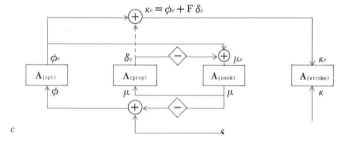

Figure 5.14 *Functional diagram of the mechanism underlying localization in mantids. The hypothesis is developed by steps from a via b to c. (a) Optic feedback loop only. As is indicated by the arrows, information flows from the optic unit (amplification factor: A*opt*) to the neck motor unit (amplification factor: A*neck*) and again to the optic unit. (b) Optic and proprioceptive feedback loops. The neck motor unit is controlled by the difference between the optic (φ*c*) and the proprioceptive (σ*c*) center messages. (c) Complete hypothesis: A*stroke* amplification factor of the central unit, which determines the direction of the stroke (κ).* [From H. Mittelstaedt, in *Recent Advances in Invertebrate Physiology.* Eugene, Oregon: University of Oregon Press, 1957, p. 59.]

are eliminated on one side and the head fixed in a turned position, the hitting errors are compounded.

Thus, the direction of the stroke is determined by feedback processes which control the position of the head (Figure 5.14). The alignment of the head preceding the strike is steered by the difference between the optic-center message (a function of the angle between prey and fixation line) and the proprioceptive-center message (a function of the angle between the head and the body axis). When fixation movements cease, the direction of the strike is set principally by the optic-center messages, and to some extent by the proprioceptive-center messages. Mantids that have been hand-fed from hatching and have never had to catch a fly are able to perform accurately at the first opportunity.

Insects as a group exhibit greater variety of feeding habits and diets than all other animals combined. Every substance imaginable is eaten by some insect. The more exotic diets include beeswax, cork, pepper, cured tobacco, poison ivy, the fur of South American sloths, and parts of the horns of African antelopes. Some insects bore into lead pipes, cables, and rubber insulation. In connection with the utilization of so many different kinds of food, diverse patterns of behavior have evolved; nevertheless, at the basis of all feeding habits are mechanisms for controlling quality and quantity.

The discrimination of quality, that is, diet selection, has been studied most extensively in plant-feeding insects. Many of these are omnivorous, but a great number are very selective, often dying of starvation when a particular plant is not available even though other species equally nutritious are present in abundance. The clues utilized in diet selection are specific chemicals or combinations of chemicals which may or may not be nutritionally important. Of the nutritionally important compounds sucrose is the most universal. The nonnutritious compounds (token stimuli for the insects) encompass a wide variety of plant chemicals (secondary substances whose metabolic roles are obscure). It is possible that they evolved as defense mechanisms on the part of the plants. Thus, an insect may feed on a particular species of plant because it contains a specific highly stimulating compound and lacks compounds that deter feeding. For example, mustard oils and their glycosides, occurring characteristically in plants of the cabbage family (Cruciferae), presumably protect Cruciferae from attack by many insects but are feeding stimulants for cabbage caterpillars. These mustard oils and glycosides have no nutritional value.

Electrophysiological recording from the taste receptors of these caterpillars has shown that there are receptors specifically sensitive to mustard oil glycosides. Further analyses have revealed, however, that the discrimination of special foods is not always mediated by special receptors which filter out all but the relevant signals. The olfactory receptors of tomato caterpillars and others respond to a wide variety of plant odors; however, each receptor has a different response spectrum so that the total input from all receptors constitutes a message which the central nervous system decodes. In these cases the basis of discrimination is central.

Under duress some caterpillars accept an alien plant and eventually may eat it readily. This acceptance indicates that although the manner in which the central nervous system decodes and acts upon sensory

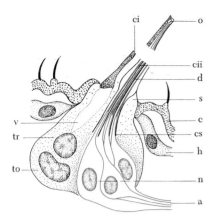

Figure 5.15   Diagram of a chemo-receptive hair of the blowfly. Three of the five neurons are shown: o, open-ing at tip; ci, thin-walled cavity; cii, thick-walled cavity; d, dendrites of chemoreceptors; s, socket of hair; c, cuticula; h, hypodermis; cs, cuticular sheath enclosing dendrites; a, axon; v, vacuole; tr, trichogen (hair-forming cell); to, tormogen (socket-forming cell); n, bipolar neurons. [From V. G. Dethier, in *Nebraska Symposium on Motivation,* D. Levine, ed. Lincoln, Nebraska: University of Nebraska Press, 1966, p. 111.]

information is genetically determined, a certain amount of plasticity does occur so that feeding habits can sometimes be modified as a result of experience.

The most thorough analysis of the quantitative regulation of feeding has been carried out on the black blowfly (*Phormia regina*). In nature its food consists of nectar from flowers, honey dew secreted on leaves by aphids, sap from wounds on trees, and many kinds of decaying material. For egg laying and ingestion of protein, flies assemble at carrion, offal, and feces. When food is odorous, the odor causes flies to switch from random flying patterns to an upwind orientation. Arriving at the food the fly steps in it. Stimulation of chemosensory hairs on the feet causes the retractable proboscis (labellum) to be extended, thus bringing marginal labellar hairs into contact with the food. Each hair houses five bipolar neurons: two salt receptors, a water receptor, a sugar receptor, and a mechanoreceptor (Figure 5.15). Whether or not the food is ingested (assuming the fly is hungry) depends on the balance between input from the sugar receptors and that from the salt receptors. The former mediate acceptance; the latter, rejection. A food that stimulates the sugar receptors more effectively than the salt receptors is accepted; a food causing the reverse effect is rejected. Ingestion by the hungry fly is com-pletely under the control of sensory input from these receptors and similar ones on the oral surface of the proboscis, which next comes into contact with the food when signals from the marginal hairs trigger un-folding of the labellar lobes. Stimulation of the oral papillae initiates sucking. It is remarkable that the entire complex process of feeding— extension of the proboscis, unfolding of the oral lobes, sucking, and swallowing—can be triggered and maintained by action potentials from a single neuron, the sugar receptor of any hair. It is also note-worthy that the effectiveness of sugars as stimuli bears no orderly rela-tion to nutritional value. One of the most highly stimulating substances is fucose, a sugar with no nutritive value whatsoever. In a choice test a fly will choose highly stimulating, nonnutritive fucose in place of mannose, a highly nutritious, poorly stimulating sugar.

As food is pumped into the esophagus by a pharyngeal pump, it is driven by peristalsis back into a blind storage sac, the crop (Figure 5.16). After a few minutes the oral sense organs adapt and ingestion ceases. By the time disadaptation has occurred, and the organs of taste are again sensitive enough to trigger ingestion, postingestion inhibitory factors have already begun to operate. Their action blocks further in-gestion.

Inhibition arises from the activity of two sets of receptors respon-sive to stretch. One set consists of two bipolar neurons in the foregut.

Periodically, after the fly has stopped feeding, food is regurgitated internally from the crop into the midgut where it is to be absorbed. In the process it passes back into the foregut before being rerouted. It is here that the two stretch receptors, bipolar neurons, are located. Action potentials generated by them pass forward to the brain via the recurrent nerve (Figure 5.16). In the brain excitatory input from the oral receptors is balanced against inhibitory input from the foregut receptors and the outcome determines whether or not feeding occurs. Another inhibiting system occurs in the form of receptors in the body wall that monitor stretch as the animal becomes full. Inhibitory information from them is also sent to the brain but via the central nerve cord. Cutting either the recurrent nerve or the ventral cord deprives the fly of information about the amount of food it has eaten. Under these circumstances food that stimulates the oral receptors is ingested until the animal finally bursts. The interactions of the mechanisms controlling feeding are summarized in Figure 5.17.

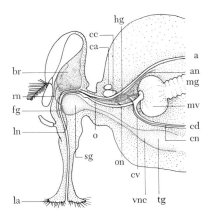

Figure 5.16 **Diagram of the head and thorax of the blowfly showing the endocrine complex and the relationship between nervous system and alimentary canal:** br, brain; rn, recurrent nerve; fg, frontal ganglion; ln, labrofrontal nerve; la, labellum; cc, corpora cardiaca; ca, corpus allatum; hg, hypocerebral ganglion; e, esophagus; sg, subesophageal ganglion; eon, nerve to esophagus; cv, crop valve; vnc, ventral nerve chord; a, aorta; an, nerve to aorta; mg, midgut; mv, midgut valve; cd, crop duct; cn, nerve to crop; tg, thoracic ganglion. [From V. G. Dethier, in *Nebraska Symposium on Motivation*, D. Levine, ed. Lincoln, Nebraska: University of Nebraska Press, 1966, p. 107.]

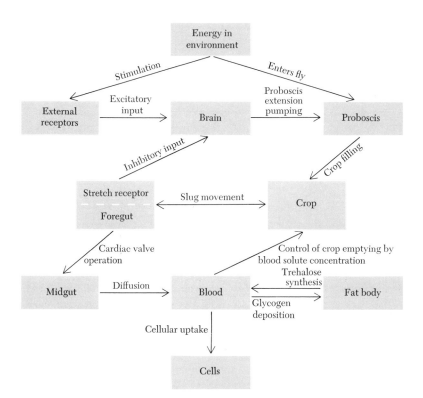

Figure 5.17 **Diagram illustrating mechanisms of metabolic homeostasis in the blowfly.** [From A. Gelperin, *J. Insect Physiol.*, *12* (1966), 829–841.]

The fly, then, possesses a relatively simple system operating by means of antagonistic sensory systems, one chemical and peripheral, the other mechanical and internal. Incoming information is integrated in the central nervous system. The system is primitive in that it is not tied in directly with nutritive values and metabolic changes; however, the fact that in nature high stimulating value and nutritive value normally are correlated and metabolic state and the gut and body wall distensions normally are correlated insures that energy needs are met efficiently.

Until the nervous system evolved to the degree of complexity found in the more complicated molluscs and the arthropods, orientation in space was largely a matter of taxes and kineses. With the development of complex eyes and the brains to process the information came the possibility of orientation by visual landmarks. In addition, celestial navigation became possible with the ability to maintain any chosen angle of progression with respect to the sun. Thus, many invertebrates and fish, reptiles, amphibia, birds, and some mammals are able to employ the sun as a compass for homing and migration. Some (amphipods) also navigate by the moon. Many invertebrates, in contrast to vertebrates, possess an additional capacity: insects, shrimps, crabs, water mites, woodlice, spiders, and octopuses are able to navigate by perceiving the plane of vibration of the polarized light in the sky. Coupling the ability to perceive the plane of polarization with a built-in biological clock, honeybees can compensate for the shift in the plane of polarization that occurs as the sun marches across the sky.

In some shore-dwelling species of arthropods as, for example, sand fleas, the direction of orientation is innate. Sand fleas displaced inland of the moist shoreline hop toward the ocean. Those living on the west coast of Italy would, under these circumstance, jump west. If they are transported to the east coast, they still jump west, that is, away from the ocean.

For most animals, however, the ability to use a sun compass is learned and a number of navigational aids are employed. Honeybees, the most diligently studied animals, employ both the sun and polarized light in their orientation. The capacity is learned. With their time sense they are able to note the time of day and the course of the sun. If exposed all their lives only to the morning sun, they are able to plot the course of the afternoon sun. They can even extrapolate to the course of the sun at night. If sun position and internal clock are experimentally put in opposition, as, for example, training bees to a certain food source on the east coast of North America and then testing them in California with its time difference, the internal clock is obeyed first, but gradually sun time replaces it.

Under many circumstances visual landmarks take precedence over the sun as navigational aids. A conspicuous forest margin is better than the sun, but individual trees or clumps of vegetation are not.

Honeybees are unique among invertebrates in that they can communicate detailed navigational information to one another. Visual landmarks are employed by bees only for their individual use. This kind of information is not communicated; however, information derived from celestial navigation is transmitted. Honeybees communicate to their hive mates the distance and direction of supplies of nectar, pollen, water, and propolis, and sites for establishing new hives. The nature of food and its relative quality and abundance are also indicated. The information is transmitted by dances and by chemical cues. There is a common dance language for all purposes, but different factors release them. The intensity of collecting and the quantity supplied is governed by the eagerness with which the substance is taken from the foragers which in turn stimulates them to increase the duration of their dances. Sample tastes inform the recipients of the nature of the substance collected. This is a nonobligatory command to fetch more of the same.

Food sources at distances shorter than 100 m are signaled by round dances made on the vertical comb (Figure 5.18). Interest in the dance is exhibited only by hive mates that were collecting from the same kind of flower. These group comrades trip around the dancing bee whose

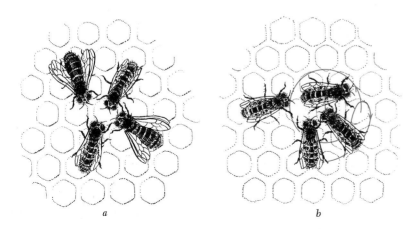

a   b

Figure 5.18 *(a)* **A forager (lower left) has returned home and is giving nectar to three other bees. (b) The round dance: The dancer is followed by three bees who trip along after her and receive the information.** [From K. von Frisch, *The Dance Language and Orientation of Bees.* Cambridge, Mass.: Harvard University Press, 1967, p. 29.]

ARTHROPODS

body is contaminated by the same flower fragrance. These then fly out in search of the same fragrance and are assisted by the odor of an abdominal scent gland everted by the collectors on the spot.

As distances of sources increase, the round dance of a forager is gradually transformed into a waggle dance, which gives information about distance and direction (Figure 5.19). The waggle dance is a figure-eight pattern in which the bee waggles her abdomen from side to side (13 to 15 waggles/sec) as she traverses the "waist" of the figure eight. A buzzing accompanies the waggle. The dance is interrupted whenever attendant bees squeak, and food is given. Several elements of the dance vary with distance: duration of a circuit, duration of return run, duration of waggling, number of waggling movements, and length of waggling segment. The *duration of waggling* is the best index of distance and decreases with increasing distance. Distances up to 11 km can be recorded. Minor variations occur in the dance, but attendant bees average a large number of waggle runs.

The estimate of distance is based on the amount of energy expended traveling the round trip, but the value assigned to the outgoing flight predominates. Excessive distances are reported when a headwind is encountered on the way out, for uphill flights, when wings are clipped, or when weights or foils increasing wind resistance are added.

When the dance is performed on a horizontal surface (as it always is with some primitive honeybees), the waggling run points directly to

Figure 5.19 **The tail-wagging dance; four followers are receiving the message.** [From K. von Frisch, *The Dance Language and Orientation of Bees.* Cambridge, Mass.: Harvard University Press, 1967, p. 57.]

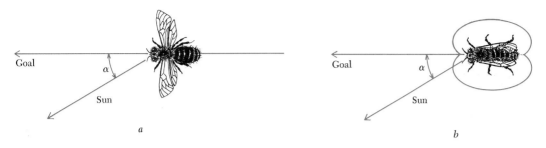

a   *b*

Figure 5.20 *Indication of direction when the dance takes place on a horizontal surface. During the dance, the bee takes the position (b) in which she sees the sun at the same angle (α) as she saw it when flying to the feeding place (a).* [From K. von Frisch, *The Dance Language and Orientation of Bees.* Cambridge, Mass.: Harvard University Press, 1967, p. 132.]

the source (Figure 5.20). When the dance is done in the dark on a vertical surface, the angle of the waggle that is run with respect to the sun or plane of polarization is transposed to a gravitational angle (Figure 5.21). If nerves to the proprioceptors employed in orienting to gravity are cut, the vertical but not the horizontal dance becomes disoriented. There is no dance language for up and down. When flying detours are necessary, the dances signal air line direction but also indicate total detour distance. The bees derive the air line direction, which they have never actually flown, by integrating the length of solar angle perceived during the different segments of flight.

In signaling potential new hive sites each exploring bee does her own dance indicating the location of her find. The more suitable the find the more vigorous the dance. Gradually the least attractive sites (dances) are ignored and the better ones are reinforced by other dances until a consensus is reached (Figure 5.22).

Of the several other methods of communciation employed by insects (for example, visual and sound) chemical communication is perhaps the most widespread. The chemicals involved, usually the products of special glands, have been given the name *pheromones* [substances secreted externally to regulate the organism's environment by influencing other animals of the same species (Figure 5.23)] Even in honeybees, with their highly effective dance language, pheromones play an important role. During her nuptial flight the queen attracts drones by scents released from her mandibular glands. Scent-fanning bees at the entrance of the hive blow its odor outward so that foragers are guided home. A bee can also give an alarm signal by extruding her sting and

Figure 5.21 *Three examples of the indication of direction on a vertical comb surface: st, beehive; I, II, III, feeding stations in three different directions; I', II', III', the corresponding tail-wagging dances on the vertical comb.* [From K. von Frisch, *The Dance Language and Orientation of Bees.* Cambridge, Mass.: Harvard University Press, 1967, p. 137.]

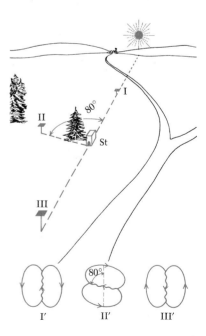

fanning to disperse an odor which stimulates the hive mates to attack.

Many species of stingless bees, which do not perform direction-indicating dances, employ pheromones instead, placing scent markers on the ground with their mandibular glands. They then lead newcomers along the trail to the goal (Figure 5.24). Among ants the technique of trail marking is widespread. The contents of different glands in various species are placed on the ground by dragging the abdomen or drawing a line with the sting. It is possible to lay an artificial trail by drawing a

*Figure 5.22 **The dances of scout bees and the achievement of agreement on one of the nesting sites discovered.** Every scout was marked when she first danced. The arrows represent, to scale, direction and distance indicated by the dances. Each arrow represents a newly marked bee. For the nesting site finally selected (300 m to ESE) the numbers of bees are indicated.* [From M. Lindauer, Z. vergleich. Physiol., 37 (1955), 263–324.]

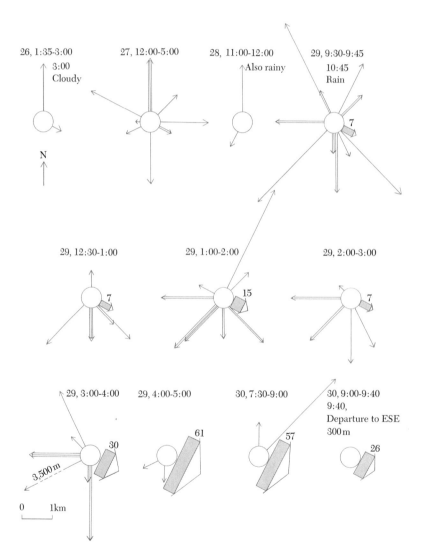

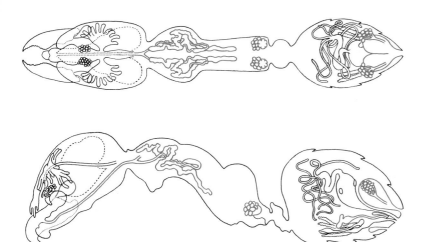

Figure 5.23   *Glands with which ants produce chemical signals (dorsal and lateral views).* [Redrawn from E. O. Wilson and W. M. Bossert, *Recent Progress in Hormone Research, 19* (1963), 673–716. After Pavan and Ronchetti.]

Figure 5.24   *Odor trails laid down on the vegetation by a forager of one of the stingless bees: solid line, first flight; dashed line, second flight; dots and circles are the spots marked with scent.* [From M. Lindauer and W. Kerr, *Z. vergleich. Physiol., 41* (1958), 405–434.]

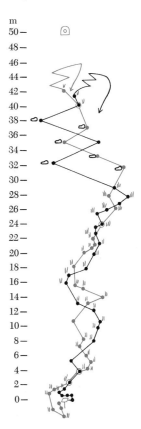

freshly dissected gland along the ground. Because of evaporation and diffusion the trail is actually a sausage-shaped area of odorous space. It is this space rather than the material actually on the ground that is detected.

Ants also produce a large number of alarm substances that stimulate either flight or attack. One such substance is citronellal, readily recognized if one crushes the appropriate species of ant. Many species of insects employ special pheromones as sex attractants to aid in bringing the sexes together and as stimuli for courtship. It is probable that pheromones play a central role in the organization of social activities of insects. Finally, insects produce a vast array of common and exotic chemicals in pure form as a means of defending themselves against predators. A striking example is the bombardier beetle, which shoots a mixture containing quinones at a temperature of 100°C. Other terrestrial arthropods also employ defensive sprays (Figure 5.25).

The wide scope of these activities leads the unwary to postulate a higher behavioral capacity than these insects actually possess. It goes without saying that habituation is a common phenomenon. Indeed, integrated social behavior can be largely explained by habituation. For example, strangers in an ant or bee nest are immediately recognized on the basis of their odor and are often killed. If, however, a stranger is introduced at a time when the colony is busy with other matters, he may pass unnoticed and eventually be accepted. One explanation is that the others become habituated to his odor.

Innate responses to stimuli also explain much of the highly organized behavior of the social insects. For instance, performance and interpretation of the intricate bee dance are not learned. Time sense is built into such insects as bees as part of the same endogenous clock system that regulates periodic activities in many animals. In navigation, the automatic correlation between the sun's position, the plane of polarization, and the pull of gravity (when the dance is being done) is innate. On the other hand, visual cues in the environment are learned.

The learning ability of insects varies greatly from one species to the next. Most insects are capable of classical conditioning; they can be made to associate an adverse stimulus such as electric shock with light, dark, and so on. Certain insects can be instrumentally conditioned. Bees can be trained to associate food with a given color or shape. Ants exhibit considerable proficiency in learning a maze (Figure 5.26).

Although learning is an important component of behavior in many ants, bees, and wasps, for insects as a class it plays a relatively minor

Figure 5.25 **Whip scorpion (Mastigoproctus giganteus)** *effecting an aimed discharge (from a pair of glands opening at rear of body) in response to pinching of one of its "feelers" with forceps. The secretion contains 84 percent acetic acid.* [Courtesy of T. Eisner.]

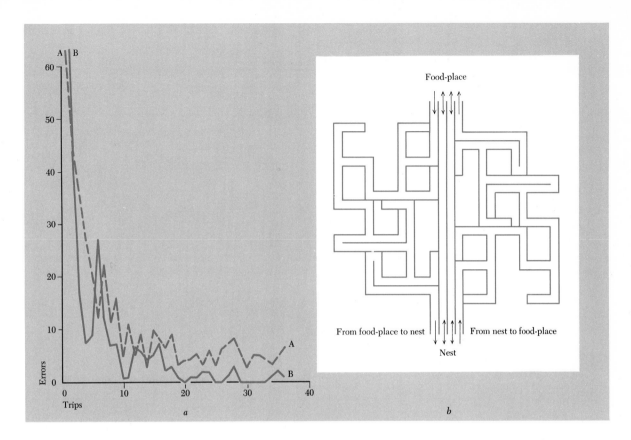

Figure 5.26 (a) *Learning curve for a fairly successful performance of Formica incerta in the maze shown in b. Line A is outward trip to food place; line B is return to nest. (b) Maze used by Schneirla (1933) in the investigation of ant learning.* [Redrawn from W. H. Thorpe, *Learning and Instinct in Animals.* Cambridge, Mass.: Harvard University Press, 1956. After T. C. Schneirla, 1933.]

role. Insect behavior may be complicated and may mimic many aspects of the behavior of vertebrates, but insects are still stimulus-bound, reflex animals. Their behavior probably represents the farthest development possible in reflex behavior. Higher learning requires further extension of the principles of multiplicity and complexity, which in turn require additional integrative units in the brain. In this endeavor, insects are defeated by their small size.

6 THE VERTEBRATE NERVOUS SYSTEM

TURNING NOW TO THE VERTEBRATES, WE COME UPON A NEW SPECIAL-
ization in the evolutionary development of the nervous system, for these
animals with backbones have a single, hollow, dorsal nerve cord that
terminates anteriorly in a large ganglionic mass, the brain. The trends
seen in the invertebrates continue in the vertebrates since there is a
further central concentration of neural tissue and an enlargement of
the overall size of the central nervous system, owing to an increase in
both the number of nerve cells and the complexity and extensiveness of
their interconnections. Most important of all, however, is the continued process of *encephalization*—the great expansion in size and functional capacities of the more anterior parts of the central nervous system,
in short, the great development of the brain itself.

The striking nature of these changes can readily be seen upon gross
examination of representative vertebrate brains (Figure 6.1). Two points
are noteworthy here. First, among the vertebrates, brain weight ranges
from a few grams in fish, amphibia, and even large reptiles to 1,200 to
1,400 g in man, and even more in large animals like the elephant and

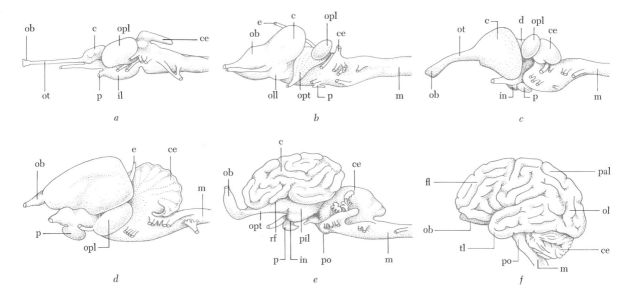

Figure 6.1 *Lateral views of representative vertebrate brains: a, codfish; b, frog; c, alligator; d, goose; e, horse; f, man. Actual sizes range from a centimeter or less in fish to 12 to 15 cm in man, and weights range from a few grams to 1,200 to 1,400: c, cerebrum; ce, cerebellum; d, diencephalon; e, epiphysis; fl, frontal lobe; il, inferior lobe; in, infundibulum; m, medulla; ol, occipital lobe; ob, olfactory bulb; oll, olfactory lobe; opl, optic lobe; ot, olfactory tract; opt, optic tract; p, pituitary; pal, parietal lobe; pl, piriform lobe; po, pons; rf, rhinal fissure; tl, temporal lobe.* [After A. S. Romer, *The Vertebrate Body*. Philadelphia: W. B. Saunders Co., 1962.]

whale. Second, there is a progressive change in the configuration of the brain, as we can see in the enormous development of the cerebral hemispheres and particularly in the cortex overlying them. In order to appreciate the full significance of the process of encephalization, it will be helpful to review briefly the structure and function of the major subdivisions of the brain and of the spinal cord leading into it.

By taking up the spinal cord first, we shall have a chance to see the organization of a relatively simple part of the central nervous system that has perhaps undergone the least change in phylogeny, and, at the same time, we shall learn something of the plan of the central nervous system. Then we can consider the major subdivisions of the brain itself and see how the great changes in behavioral capacities from fish to mammals, including man, can be related to the morphological changes in the nervous system that occur in encephalization.

In all vertebrates, the spinal cord has two major functions. One is the *integration of reflex behavior* occurring in the trunk and limbs, and the other is the *conduction of nervous impulses* to and from the brain. A reflex is a simple response to a simple stimulus, such as the knee jerk or the withdrawal of a limb from painful stimulation. The basic spinal mechanism for integrating reflexes is the *reflex arc* (Figure 6.2). Typically, it consists of the following components: (1) *receptors* in the skin, muscles, and joints that are selectively sensitive to various stimuli and form the beginning of (2) *afferent* (or *sensory*) *neurons* that enter the dorsal part of the spinal cord and terminate in contact with (3) *interneurons* (*associational neurons*) that in turn terminate upon (4) *efferent* (*motor*) *neurons* that pass out of the ventral part of the cord to end in appropriate effectors, which may be either muscles or glands (Figure 6.2).

In some cases, such as the knee jerk, the sensory neurons connect directly with the motor neurons in a *monosynaptic arc* reminiscent of the direct sensory-motor connections found in the simpler invertebrates. As a rule, however, it is through the many connections of the associational neurons within the spinal cord that information coming over various sensory neurons is organized and integrated so that a pattern of activity is set up in the motor neurons, leading to an organized pattern of response. This response is a spinal reflex such as that seen in the withdrawal or flexion of a limb when a noxious or painful stimulus is applied to its distal end. While such a reflex looks simple, it is actually a complex, patterned response, requiring the integrated action of many muscles and involving a number of basic neurophysiological mechanisms. For example, in order to execute a *flexion reflex,* an organism must contract its flexor muscles, pulling the limb toward the body, and simultaneously relax the antagonistic extensor muscles. This pattern of stopping an activity when its antagonistic activity is started is the classical pattern of *reciprocal excitation* and *inhibition* that we will see many times in the study of the nervous system. At the same time that one limb is flexed, it may be necessary for the animal to extend its opposite limb for support against the ground in a *crossed-extension reflex,* which calls for a contraction of extensor muscles and a relaxation of flexors. Together these two reflexes make up the basic pattern of stepping.

In all these actions, the contracting muscles may actually stretch the opposing relaxing muscles. Whenever muscles are stretched by the action of their antagonists, stretch receptors embedded in the muscles are acti-

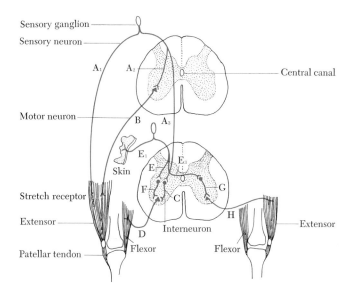

Sensory ganglion
Sensory neuron
A₁ A₂
Central canal
Motor neuron
B   A₃
E₁
E₂ E₃
Skin
Stretch receptor
F   G
C
H
Extensor
Extensor
Interneuron
D
Patellar tendon
Flexor
Flexor

Figure 6.2   Schematic diagram of two adjacent segments of the spinal cord, showing examples of connections of sensory neurons, interneurons, and motor neurons making up various reflex arcs. The monosynaptic arc involved in the knee jerk is activated when the extensor muscle is stretched by a patellar tap. This involves sensory neuron A₁ leading from the stretch receptor through its central branch A₂ to the motor neuron B, which activates contractions in the extensor, lifting the leg. At the same time, there is reciprocal inhibition of the antagonistic flexor muscles, probably through the polysynaptic arc, involving another branch of the sensory neuron A₃ and the interneuron C, resulting in inhibition of the motor neuron D and relaxation of the flexors.

The polysynaptic arc in the withdrawal reflex to painful stimulation of the skin on the distal part of a leg involves sensory neuron E₁ and its central branch E₂, interneuron F, and the excitation of motor neuron D, leading to a contraction of the flexors.

At the same time, a crossed-extension reflex occurs over another branch of the sensory neuron E₃, through the interneuron G and the motor neuron H, resulting in excitation of the extensor muscles on the opposite side. Not shown in these last two cases are the reciprocal inhibitory influences on the muscles antagonistic to those contracting and the feedback from the stretch receptors that are present in all the muscles. Also, it should be pointed out that while only one neuron is portrayed in each component of the reflex arc, actually many, many neurons typically work in concert. [Modified from D. P. C. Lloyd, "Synaptic Mechanisms," in A Textbook of Physiology, J. F. Fulton, ed. Philadelphia: W. B. Saunders Co., 1955.]

vated, and a *stretch reflex* is initiated. In this reflex, the stretched muscles begin to contract and the antagonistic muscles begin to relax, offsetting the original reflex somewhat. In addition, the stretch reflex activates special small motor neurons that go back to the very same stretch receptors, and their action there may result in a reduction of the stretch imposed on the receptors so that they are "reset" to an optimal range of sensitivity. Then they can more readily respond to still further stretch that may be imposed upon them. The net effect of all these neurophysiological mechanisms is to permit movements to be executed in a smooth and graded fashion, sensitively integrated into a meaningful whole with all other movements.

All these reflexes may be obtained in the spinal animal where the spinal cord is severed from the brain, and indeed the basic integrating mechanisms may be contained within a few segments of the spinal cord. So it is easy to see, in even a simple system, what an important role the central nervous system may play in the sorting of sensory information and in the organization and patterning of motor responses. More complex reflex functions like the *scratch reflex* and *walking reflexes* require the organization of even more sensory information over many segments of the spinal cord and the coordination of even more complex patterns of response.

In addition to mediating skeletal muscle reflexes, the spinal cord is connected, through special nerves branching off from it, to the *autonomic*

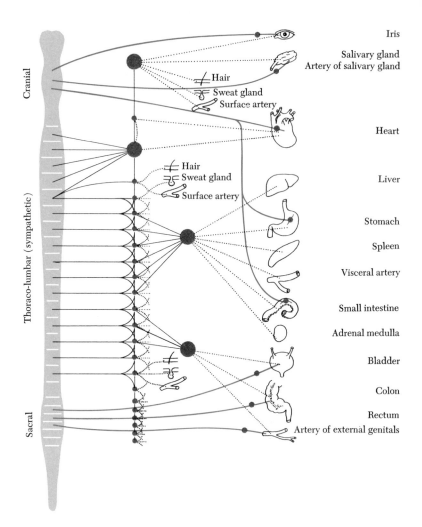

Iris

Salivary gland
Artery of salivary gland

Hair
Sweat gland
Surface artery

Heart

Hair
Sweat gland
Surface artery

Liver

Stomach

Spleen

Visceral artery

Small intestine

Adrenal medulla

Bladder

Colon

Rectum

Artery of external genitals

Cranial

Thoraco-lumbar (sympathetic)

Sacral

Figure 6.3 **Diagram of the vertebrate brain and spinal cord (left), showing the sympathetic ganglionic chain in the thoracic-lumbar region of the cord and the parasympathetic outflow from the brain and the sacral region of the cord.** [Redrawn from W. B. Cannon, *The Wisdom of the Body*. New York: W. W. Norton, 1932.]

nervous system, which innervates the viscera, blood vessels, and other smooth muscles of the body. As you can see in Figure 6.3., the *sympathetic branch* of the peripheral autonomic nervous system, lying along the middle region of the cord, consists of a chain of ganglia that sends out a diffuse network of fibers to the organs innervated. From the hindmost region of the cord and from the brain arise the nerve fibers of the *parasympathetic branch* of the autonomic nervous system; these fibers go directly to the individual organs they innervate over discrete, individual pathways. In general, the sympathetic system functions in active processes that expend energy (increased heart rate, blood pressure, pupil dilation, and so on) and the parasympathetic system functions in repair (promotes digestion, sleep, and so on), but this "antagonism" is not always distinct or consistent. Both systems have their central representation in the various parts of the brain that are concerned with visceral functions (medulla, hypothalamus, old cortex).

When the influence of the brain on spinal reflex mechanisms is intact, the complexity of reflexes and of other behavior increases vastly. This is achieved by nervous conduction to and from the brain through long columns or tracts of nerve fibers (axons) that run up and down the peripheral parts of the spinal cord. Upon gross examination, the columns appear white (Figure 6.2) because the larger axons in them are sheathed in a white fatty substance called myelin. The butterfly-shaped core of the spinal cord is gray since it consists mainly of cell bodies of associational and motor neurons that are unmyelinated. The *ascending (sensory) columns* consist mainly of long branches of peripheral sensory neurons coming into the spinal cord and crossing to the opposite side early in their central course (Figure 6.4). In a large animal, these individual nerve cells may thus be many feet in length, especially in the case of peripheral nerves arising in the foot and going up into the brain. To a large degree, different sensory systems are segregated within the spinal cord so that there are somewhat separate columns mediating pain,

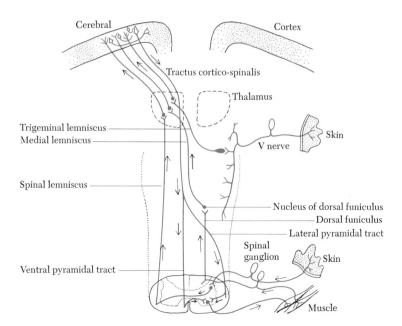

Figure 6.4 **Diagram of the ascending and descending pathways between the spinal cord and the brain, showing the crossing of pathways from one side of the nervous system to the other.** [Redrawn from C. J. Herrick, *An Introduction to Neurology.* Philadelphia: W. B. Saunders Co., 1931.]

temperature sense, muscle sense (or proprioception), and touch. The *descending (motor) columns* consist of neurons going all the way down from the brain and crossing to the motor nerve cells in ventral parts of the spinal cord (Figure 6.4). Because of this crossing in both the motor and sensory neurons, the left side of the brain typically controls the right side of the body and vice versa, so that after brain injury there is usually paralysis or loss of sensitivity on the opposite side of the body.

In the course of vertebrate phylogeny, the basic organization of the spinal cord underwent relatively little change. As the forward parts of the brain developed, however, the ascending spinal columns extended up to them (thalamus), and the descending columns originated in the highest parts of the cerebrum, the cortex itself. At the same time, the tracts became more concentrated and distinct and their functional connections more discrete, all of which permitted greater refinement and segregation of functions.

In this discussion of the spinal cord, we have illustrated in somewhat simple form some of the basic principles of organization and function that apply to the vertebrate nervous system as a whole. First, as was illustrated in the reflex arc, we can see that the nervous system, with its sensory, associational, and motor components, is a great stimulus-response correlating mechanism. Second, it operates according to a principle of reciprocal excitation and inhibition so that as one activity is started, another—perhaps antagonistic—activity is actively stopped. Third, it has built into it a feedback mechanism from the muscles so that it is promptly influenced by the consequences of its own activities. Fourth, in addition to such sensory monitoring of motor activities, there is also a motor or efferent modulation of sensory activity, as in the case of "resetting" of the stretch receptor.

As we turn to the complexities of the brain itself, we shall see additional principles of nervous action illustrated, for the brain has additional association mechanisms for the integration of sensory and motor activities over its own neurons or, through the ascending and descending columns, over the spinal neurons. Thus, in the case of man, for example, a painful stimulus in the toe results automatically in a spinal reflex of withdrawal; in addition, however, the information is integrated at several levels within the brain, a sensation and perhaps an emotion result, and the reflex may be modified or inhibited in the light of other information reaching the brain—by attitudes, past experiences, and so on. Not only is the reflex behavior mediated through the vertebrate brain likely to be more complex and variable than are spinal reflexes, but, as we shall see, additional behavioral properties also appear as a result of both the enormously complex network of associational neurons and the specialized sensory and motor mechanisms of the brain.

The brain itself may be divided into three major regions—the forebrain, the midbrain, and the hindbrain—that appear early in the embryonic development of all vertebrates as three primary vesicles, or swellings, in the anterior end of the neural tube. As Figure 6.5 shows, the anterior and posterior regions of the brain further subdivide so that the typical vertebrate brain has five divisions in adulthood; each division has its own specialized structures, and each structure makes its own contribution to function. As we examine the changes that occur in the evolution of the vertebrate brain, a number of points should be kept in mind.

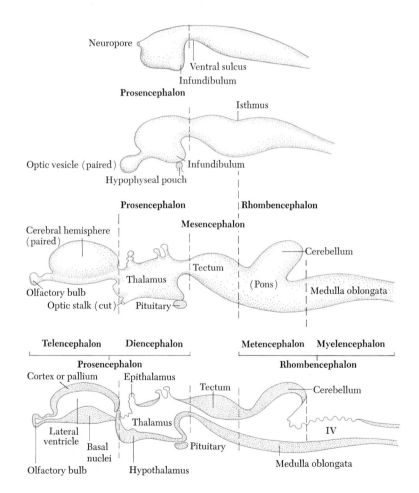

Figure 6.5 *Development of the main divisions of the vertebrate brain from the three primary embryonic vesicles; the forebrain (prosencephalon), midbrain (mesencephalon), and hindbrain (rhombencephalon).* [After A. S. Romer, *The Vertebrate Body.* Philadelphia: W. B. Saunders Co., 1962.]

1   Like the spinal cord, the brain may be divided into sensory and motor structures with great association areas in between.

2   To a large degree, the major evolutionary developments of the brain were occasioned by the development of special receptors whose nerve fibers terminated in different parts of the primitive brain. As the receptors for posture, balance, and equilibrium became more important, a special area in the hindbrain enlarged, becoming the cerebellum. The dorsal part of the midbrain, which was the original terminus of the visual pathways, greatly increased in size as the visual receptors developed. Finally, the cerberal cortex of the forebrain appeared first as the region of the brain mediating olfactory functions.

3   As the more anterior parts of the brain developed and enlarged, they mediated functions once carried out solely by more posterior structures. The result is that a given class of functions in higher vertebrates may be carried out at several levels of the brain, the anterior levels mediating the more complicated and more recently evolved of these functions and the posterior levels the simpler and more stereotyped.

4   Not all the changes we can trace in vertebrate phylogeny, however, are simple linear increases in the size and functional importance of given structures, for many of the living animals we examine are highly specialized, with highly specific adaptations in structure and function. For example, in certain fish that have taste receptors all over the body (catfish), there is enormous development of the vagal lobes of the medulla, where the vagus and other taste nerves have their cell bodies. On the other hand, the taste receptors in mammals are confined to the tongue, and these regions of the medulla are relatively small. With these points in mind, we can now examine the five major divisions of the brain.

MYELENCEPHALON   The hindmost part of the brain, consisting of the *medulla*, is essentially an extension of the spinal cord into the brain, for it contains the continuation of the sensory and motor columns of the cord. In addition, like the cord, it has its own dorsal and ventral sensory and motor nerves entering and leaving it; these are the cranial nerves that serve the skin and muscles of the head and the special sense organs concerned with taste, hearing, and balance. Throughout the vertebrates, the medulla functions in the control of vegetative functions, particularly respiratory and cardiovascular ones, but in higher vertebrates many of these functions are overshadowed by the development of vegetative functions in the hypothalamus and the older parts of the cerebral cortex.

METENCEPHALON   The most prominent structure in this anterior region of the hindbrain is the *cerebellum*, a complex structure serving in the coordination of movements and the maintenance of posture, tone, and bodily equilibrium. In early vertebrates, the cerebellum developed as an enlargement of the vestibular and lateral-line centers and received sensory fibers from the muscles (proprioception). Later, new parts of the cerebellum called the cerebellar hemispheres developed, and they received fibers from all the sensory systems.

MESENCEPHALON   This is the midbrain. Its dorsal sensory centers, making up the *tectum,* developed originally in association with optic fibers coming into this region of the brain and became the primary visual center, the large optic lobe, in fish and amphibia. As visual centers in the thalamus and cortex developed ahead of it, the midbrain tectal region, now differentiated as the *superior colliculus,* remained a center for simpler visual functions, off the mainstream of pathways going from the eye to the thalamus and cortex. At the same time, just behind the superior colliculus, the *inferior colliculus* developed as one of several auditory centers relaying to the thalamus and cortex. The lateral and ventral parts of the midbrain, the *tegmentum,* contain the main ascending and descending columns and in addition have local motor reflex mechanisms.

DIENCEPHALON   Within this division of the forebrain are located the *thalamus* in the dorsal region and the *hypothalamus* beneath it. Ahead of the hypothalamus is the entrance and crossing point of the *optic nerves,* and below it is the *pituitary gland.* The dorsal part of the thalamus is the termination point of many of the ascending columns arising in the spinal cord and medualla and, as such, is the great sensory integration mechanism in the lower vertebrates. As the cerebrum develops, certain parts of the thalamus become intimately connected with the sensory regions of the new parts of the cortex and relay information to them. The hypothalamus, as we have already mentioned, has vegetative functions and is concerned with the control of body temperature, sleep, feeding, water balance and drinking, emotion, and reproductive behavior. Also, it exerts neural and humoral control over the pituitary gland, which, besides its own endocrine functions, controls the functions of other endocrine glands of the body.

TELENCEPHALON   The most anterior division of the brain is made up mainly of *olfactory centers,* the *basal ganglia,* and the *cerebral cortex.* In the lower vertebrates, the cerebrum developed in association with the olfactory bulbs, and there is a progressive increase in the relative size of the bulbs and associated cerebral olfactory centers through the infra-primate mammals. In the primates, however, the olfactory bulbs are relatively small, and the additional development of the cerebrum is unrelated to smell. The basal nuclei of lower vertebrates function as the highest motor integration mechanism, but in the mammals, the most complex motor functions are carried out by the cerebral cortex, particularly the motor cortex. The cerebral cortex is a gray mantle, or bark,

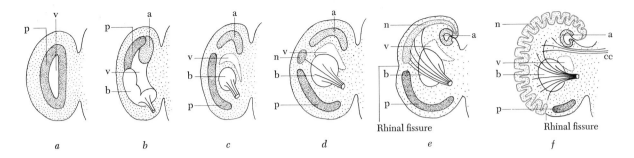

a   b   c   d   e   f

Figure 6.6   *Diagram showing the development of the cerebral cortex in the vertebrates. At first, there is a collection of cell bodies deep within the hemispheres (a and b). As the early reptilian stage is reached (c), the cell bodies appear on the surface as a mantle or true cortex. This is old cortex, divided into paleopallium ventrally and archipallium dorsally. In later reptilian stages, the neocortex appears as a small area between the two regions of old cortex (d). In early mammals, the neocortex enlarges (e), and in more advanced mammals (f), this enlargement of neocortex is so great that it not only occupies the whole dorsal and lateral surfaces of the cerebrum, but it also must fold in on itself to fit into the confines of the cranial vault. Note that the archipallium has folded into the internal mass of the cerebrum dorsally (e and f) and that the paleopallium is now a small ventral and medial part of the cortex in (f):* a, archipallium; b, basal nuclei; cc, corpus callosum; n, neopallium; p, paleopallium; v, ventricle. [Redrawn from A. S. Romer, *The Vertebrate Body.* Philadelphia: W. B. Saunders Co., 1962.]

Figure 6.7   *Lateral view of the human cerebral cortex, showing the major division into lobes.* [Redrawn from S. W. Ranson and S. L. Clark, *The Anatomy of the Nervous System.* Philadelphia: W. B. Saunders Co., 1959.]

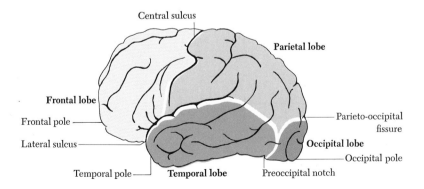

of nerve-cell bodies that overlies the cerebrum. Its development is illustrated in Figure 6.6. The oldest parts of the cortex (paleopallium and archipallium) are largely olfactory and visceral in function. The new cortex, or *neocortex,* appears in the reptiles as a receiving area for sensory fibers from lower regions of the brain. As it expands in size in the mammalian series, two things happen: (1) the paleocortex moves to a more ventral position and the archipallium, now the hippocampus, folds into the cerebral mass underneath the neocortex; (2) the neocortex expands so much that it begins to fold in on itself, developing the fissures, or sulci, and the bulges, or gyri, that characterize the convoluted brain. Using the major fissures as landmarks, we can divide the cortex into major lobes, as Figure 6.7 shows for man.

Early in mammalian evolution, the neocortex is primarily a sensory receiving area (Figure 6.8). Thus, the rabbit and the rat, whose cortices are convoluted very little, have fairly well organized and somewhat segregated areas for vision, hearing, somatic sensation, and taste that occupy most of their cortical surfaces. The motor area, anterior in the cortex, is relatively small and not highly organized, and there are very few and small areas in the cortex that are neither sensory nor motor, but are associational in function. In the carnivores (cat), the amount of associational cortex is increased, but the real increase in associational cortex occurs in the primates, especially in man (Figure 6.9), in whom only relatively small strips of cortical tissue are devoted to sensory and

Figure 6.8   Maps of the sensory and motor areas of the cerebral cortex of representative mammals. Note that as one goes higher on the scale, there is relatively less cortex devoted to sensory and motor functions and more to associational functions (compare Figure 6.9). [Redrawn from J. E. Rose and C. N. Woolsey, *EEG Clin. Neurophysiol., 1* (1949), 391.]

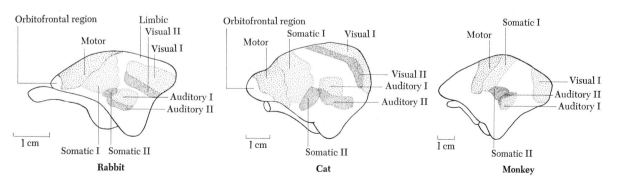

Figure 6.9 *Lateral view of the cortex of man, showing the relatively small areas devoted to sensory and motor function and the relatively large areas devoted to associational function (compare Figure 6.8).* [Redrawn from G. J. Romanes, ed., *Cunningham's Manual of Practical Anatomy*, 13th ed. London and New York: Oxford University Press, 1967, Vol. III, p. 248.]

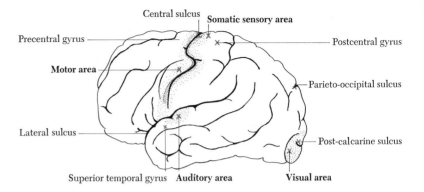

Figure 6.10 *(a) Diagram of section cut through a human brain, showing the representation of body parts as measured by evoked potentials recorded from each point of the cortex upon stimulation of the skin at the parts of the body shown. (b) Diagram of a section cut through a human brain, showing the representation of parts of the body on the motor cortex as points eliciting movements, when electrically stimulated, of the parts of the body shown.* [After W. Penfield and T. Rasmussen, *The Cerebral Cortex of Man.* New York: Crowell Collier and Macmillan, Inc., 1952.]

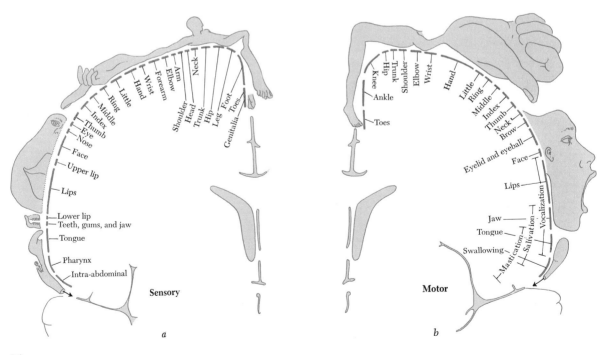

motor functions, and most of the cortex to large association areas.

Despite their relatively small size, the sensory and motor areas of the human cortex become highly organized and represent, in great detail, the functional parts of the body and the receptor surfaces. For example, electrical stimulation of the human cortex reveals that simple, discrete movements in localized regions of the body can be evoked by stimulation of discrete points on the cortex. In fact, a map can be made of the motor area, showing the orderly representation of body parts, laid out in the pattern of a homunculus (Figure 6.10b). A similar map or sensory homunculus (Figure 6.10a) can be made of the somatic sensory area by stimulating receptors in the skin and recording the evoked potentials in the cortical area. As the figures show, the amount of cortex devoted to a given part of the body is related to its functional importance, particularly its sensitivity and the fineness of its motor control. Other sensory receiving areas of the cortex are similarly organized so that the receptor surface is represented spatially on the cortex. In fact, each sensory system is represented in the cortex in at least two closely related, but separate, areas, each organized to represent the receptor surface.

As we shall see later, destruction of these separate cortical sensory areas results in defects in sensory capacity, but in examining the mammalian series, we find that the sensory functions appear "corticalized" to different degrees. Thus, destruction of the visual area in monkey and man results in virtually total blindness; in contrast, visual cortex damage in the cat and rat, while impairing their ability to see patterns, leaves them still quite capable of telling light from dark, responding to movement, and so on. Even more striking evidence for corticalization in the mammalian series is seen with destruction of the motor areas, for an animal like the rat shows essentially no obvious motor defects after such an operation, whereas the limbs of monkeys and men are badly paralzed on the opposite side of the body following destruction of one side of the motor area. Cats and dogs show some paralysis and poverty of movement, but show much more recovery of function than do the primates. It takes very careful and somewhat elaborate testing to see the effects of destruction of associational cortex, and usually such defects show up only in man and monkeys, as we shall see later.

Thus far in our effort to portray the major changes that have occurred in the evolutionary development of the vertebrate nervous system, we have revealed only the barest outline of its basic plan. The key to this plan lies, as it does in the invertebrate nervous system, in the concept of the reflex arc, with its sensory, motor, and associational neurons. Through

THE VERTEBRATE NERVOUS SYSTEM

the great ascending and descending columns of the spinal cord, this sensory-motor arc is extended into the brain, where, at various levels, additional sensory-motor integrations occur through even more complex networks of associational neurons. Let us look more closely at the organization of this system, taking the primate as our example.

On the sensory side, the great afferent systems are organized anatomically and physiologically in such a way that they reveal within the central nervous system a detailed representation of sensory qualities and spatial relationships established at the peripheral receptor surfaces. In representing sensory qualities, the afferent systems are said to be *modality-specific* in two respects. First, the different senses themselves are anatomically segregated into different parts of the nervous system representing vision, hearing, smell, taste, and the skin and muscle senses. Second, within a given sense, as we know from the study of both man and animals, there is a segregation of such sensory qualities as touch, warmth, cold, deep pressure, and pain in the skin, or the different pitches of sound from high to low. In addition, because spatial relationships established at the peripheral receptor are also preserved anatomically and physiologically within the central nervous system, they are said to be *somatotopically* organized, as the homunculus in Figure 6.10*a* and *b* shows. Not only is this somatotopic arrangement apparent in the cortex, but, as one might expect, it is preserved to one degree or another at each level of the sensory relay systems below the cerebral cortex.

At the same time that each sensory system is kept highly segregated from the others and highly differentiated within itself, it also makes a fairly nonspecific contribution to a diffuse, sensory arousal system, located in the midline regions of the brain. Each ascending column, early in its course, gives off collateral nerve connections to the core of the brain; this is known as the *reticular system,* and does not seem to preserve receptor-surface locus or sensory modality. The reticular system is literally a feltwork of intermingling and interconnecting nerves that runs on up to the midline region of the thalamus, and from there to all parts of the cerebral cortex and, indeed, to many other parts of the brain as well. Its function seems to be generally to activate or arouse the brain upon receipt of sensory stimuli at the peripheral sense organs. Destruction of this reticular system results in coma or deep sleep, even though the main ascending columns are left intact. In this condition, an organism can transmit sensory information to its cerebral cortex along its ascending columns, but cannot be aroused to act upon it.

On the motor side, there are also two major systems. One is the long, discrete *pyramidal system* going from the motor cortex to the motor nerves of the spinal cord. This system, as we have described, is highly specific and localized, for electrical stimulation of the motor cortex results in discrete localized movements of limited parts of the body (Figure 6.10b). Also, damage to it results in a loss of refined and skilled movements of the hands and feet. The second motor system is the *extrapyramidal system* arising from different parts of the cortex and various subcortical motor centers as well, including the cerebellum. Its pathway is made up of short neurons that make connections at many levels of the brain before reaching the motor nerve cells of the spinal cord. In simple terms, the extrapyramidal system functions to maintain an organized background of posture and muscle tone against which the discrete, localized movements mediated by the pyramidal paths can occur.

Sensory and motor systems are not by any means wholly separated from each other, however. Motor influences from the brain reach down into the sensory systems at many levels and modulate the activity coming into the central nervous system. In some cases, motor neurons go directly to the receptors themselves (for example, stretch receptors in muscles). In other cases, they send branches to connect with the sensory paths at some relay station within the brain. The sensory systems, on the other hand, are continually active in bringing back to the central nervous system information on the effects of motor activity by means of feedback loops, so that subsequent motor output may be modulated by the sensory effects of previous motor activity.

Associational systems serve to integrate the sensory activity coming into the nervous system over many different sensory nerves and to impose an organized pattern on the motor output, thus playing a most important role in the organization of behavior. Presumably, this activity of the associational system is determined by its given structure, the effects of past experience, motivation, and attention, as well as by the direct consequence of sensory input; further, in man, the associational systems are probably responsible for that intrinsic pattern of activity of the central nervous system known as thinking.

So far we have only discussed what has been called the somatic nervous system, which is concerned with perception, response of the skeletal muscles, and overt behavior. There is also the *autonomic nervous system,* which regulates visceral and emotional functions. This is a system interwoven with the somatic system at all levels. At the spinal level, it is represented by two sets of peripheral ganglia, the sympathetic and parasympathetic systems, both of which innervate the smooth muscles of the gut and blood vessels and many of the endocrine glands. In the brain, important autonomic functions are integrated in the medulla, the hypothalamus, and the older parts of the cortex known as the rhinencephalon.

We can now summarize the major principles of action of the vertebrate nervous system.

1 It is a vast *correlation system* for receiving sensory information, integrating it, and associating it with appropriate patterns of motor responses.

2 Its *sensory systems* serve to carry faithfully reproduced information to the brain, and at the same time, through the reticular activating system, they serve to keep the brain active and the organism alert.

3 As we saw in the spinal cord, the nervous system acts according to a principle of *reciprocal pattern of excitation and inhibition* so that as one activity is started, another antagonistic activity is slowed or stopped.

4 Its *motor systems* serve to maintain a background of posture and tone against which discrete, phasic movements are executed.

5 The sensory and motor systems mutually modulate each other through motor control of sensory input and sensory feedback of the effects of motor activity, allowing the greatest degree of refinement of both sensitivity and movement.

6 Its *associational systems* serve to correlate the various sensory influences coming into the brain and to integrate them with past experience so as to impose an organized pattern of excitatory and inhibitory activity upon the execution of behavior over the motor pathways.

7 Its *autonomic system,* organized in a parallel sensory-motor plan, serves to control visceral and endocrine activities and emotional and motivated behavior.

8 Despite this artificial separation into systems, the *action of the nervous system is integrated* into an organized whole serving the behavioral adaptations of the organism.

9 There is in this organized whole a *hierarchical order of integrations,* with reduplications of functional mechanisms at various levels of the nervous system.

10 In keeping with the evolutionary process of *encephalization,* the more complex integrations are mediated by the more anterior and more recently evolved neural mechanisms.

UP TILL NOW, WE HAVE BEEN MAINLY CONCERNED WITH THE BASIC biological machinery for behavior. We have discussed the evolution of the nervous system through the invertebrates and the vertebrate series, showing how the complexity of the nervous system has grown in phylogeny and pointing out how this has made possible the great increase in the refinement, variability, and complexity of behavior. We have also discussed the physiology of neurons and networks of neurons to give you some understanding of the "language" of the nervous system, the means by which the multitudinous parts of the nervous system communicate with one another, and how they receive information from the external world, encode it, integrate it, store it, and translate it into behavioral action. All along, we have emphasized that it is out of these anatomical and physiological properties of nervous tissue that behavior, as we observe it in organisms, is possible.

Now we come to a consideration of behavior itself and how it serves the adaptive ends of the individual organism and the species. Here we shall divide animal behavior into meaningful units, analyze it, and see in some detail how it is dependent on the functioning of the nervous system. In making such an analysis of behavior, we must recognize that there is, in the phylogenetic series, an enormous range of complexity of behavior, from the simple, brief, stereotyped act to the highly intricate and highly variable long sequence of acts.

At first in phylogeny, behavior is largely a matter of a stimulus triggering a response or of a pattern of stimuli triggering a sequence of responses. At this point, behavior is *stereotyped* and the organism is to a large extent *stimulus-bound*. Since this kind of behavior is essentially the outcome of the inherited properties of the nervous system of the organism or species in question, we speak of it as *innate*. Later, behavior becomes more variable and, what is particularly important for us, more modifiable through experience. The adaptations of an individual organism may develop uniquely in its life history through the process of learning, and we speak of such behavior patterns as being *acquired*. In the simpler organisms, what is acquired may be fairly simple and still pretty much stimulus-bound. But as the complexity of the nervous system grows and as we examine the existing species of mammals, especially the primates and man himself, still new properties of behavior emerge. Behavior is now not so stimulus-bound; much of it may originate within the organism on the basis of past experience; and much of it may be guided by complex symbolic processes such as language in man. Such intrinsic processes constitute *reasoning* in man, and we see the rudiments of it in animal behavior serving the adaptive ends of the organism.

In this and the remaining chapters, we shall take up these various modes of adaptive behavior. First, we shall consider the major stereotyped, innate modes of adaptation, the *taxes* (direct orientations of the organism with respect to stimuli), the *reflexes*, and the *instincts*. In Chapter 8, we shall take up *learning ability* and *problem solution* or *reasoning*, which are the more variable, acquired modes of adaptation. In Chapter 9, we shall address the question of how individual organisms relate to each other through taxes, instincts, learning, reasoning, and symbolic processes to produce *social organizations* that obviously serve very important adaptive functions.

As you will see, the importance of these particular modes of adaptation to an organism changes in the course of phylogeny (Figure 7.1). In man, the dominant modes of adaptation are reasoning and learning, and there is very little in the way of instinct or even reflex that is not greatly modified by experience; taxes are essentially nonexistent. In a simple mammal like the rat, on the other hand, reasoning is virtually nonexistent, but learning is well developed; instincts clearly are present and important, but they may be modified by experience; some taxes are present, but only very early in ontogenetic development. The insects are, relatively speaking, poorer learners, are dominated by largely un-

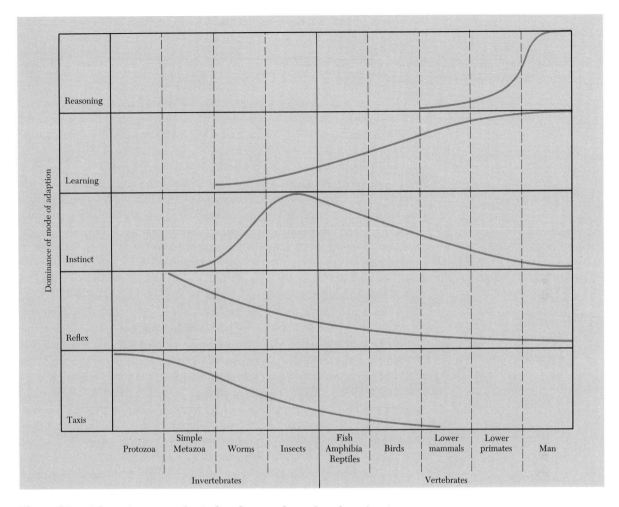

Figure 7.1   Schematic portrayal of the changes that take place in the major modes of adaptive behavior in phylogeny. Reading from left to right, it shows the relative development of different modes of adaptation; reading up and down, it shows the relative pattern of modes of adaptation at different levels of the phylogenetic scale.

modifiable instincts, and show taxes quite clearly. Below the level of worms, learning is not clearly recognizable and may not be a property of organisms; instinctive patterns are relatively simple and poorly developed, and organisms are dominated by taxes and reflexes.

Although attempts to portray trends in the major modes of adaptation in the course of evolution are necessarily speculative, especially in view of the great gaps in our knowledge, several important principles are apparent. (1) As complex modes of behavior become possible in phylogeny, they are at first added to the simpler modes, but eventually replace them. (2) At each new level of complexity, new behavioral properties emerge. (3) It is fallacious to try to account for the behavior of relatively simple organisms in terms of human capacities, such as reasoning (Lloyd Morgan's Canon). (4) It is equally fallacious to try to account for the behavior of relatively complex organisms, such as the lower primates and man, primarily in terms of simple stimulus-bound relations, instincts, or blind trial-and-error learning. The primates have new modes of adaptation that overshadow the simpler ones, and we need a converse of Lloyd Morgan's Canon that will warn against underestimating the capacity of animals with complex nervous systems on the basis of principles discovered in the study of simpler organisms.

TAXES AND ORIENTATION

Perhaps the simplest form of adaptive behavior is the orientation of an organism to some aspect of its environment. Not all orientations are taxes, however. In a very simple case, the orientation might be nothing more than a series of random movements coupled with occasional avoidance or approach movements in response to a specific stimulus. Thus, paramecia will congregate in a region of low CO_2 concentration. Whenever random movements bring them close to the bubble where CO_2 concentration is high, they swim backwards, turn, and then swim forward again, away from the bubble. This action is repeated over and over again, with the result that most of the paramecia of a group are situated at some distance from the bubble at any given instant.

This kind of orientation does not qualify as a taxis, for its direction is not continuously guided by a specific stimulus. An example of a taxis would be orientation to maintain equal stimulation of two bilaterally symmetrical receptors or, by alternate left and right movements, equalization of stimulation at successive intervals over time. For example, an organism may orient toward a light source so that both eyes receive equal stimulation. If the source is moved laterally, the orientation will change because one eye is now receiving more illumination than the other. If one eye is removed or painted over, the organism will move continuously in circles, as though trying to equalize the light on the two eyes. Such an orientation, continuously and specifically guided by external stimuli, is called a *taxis*. Jacques Loeb, a pioneer in the study of this behavior, called this a tropism, but modern students reserve *tropism* for the orientation of plants by growth (for example, toward light, away from gravity).

In some simple cases, the orientation is not guided continuously by external stimuli, for it may be a very brief all-or-none affair, such as the rapid strike of the mantis in capturing prey. Again, the orientation may occur after the stimulus is over, as in the male firefly's response to the brief flash of the female. On the other hand, an orientation may involve more than a simple taxis, even though the organism is continuously guided by the external stimulus and the consequences of its own responses to it. The optomotor turning response in the orientation of the beetle (*Chlorophanus*) to the movement of light is an interesting example. In an ingenious test situation, the beetle is fixed in position inside a drum on which rotating lines of light can be projected. Its feet grasp a very light-weight Y-maze globe, and as the direction of the rotation is changed from left to right, the beetle's locomotory movements take him along the left or right choices of the Y maze (Figure 7.2). The direction and speed of the beetle's orientation response to the rotating lines is a function of the direction and speed of the rotation and the size and brightness of the lines against the background drum. The operation of these variables is so precise that it can be expressed in the mathematical form used in the study of control systems in engineering

The adaptive value of taxes can be seen in the case of the grayling butterfly (*Eumenis semele*), which flies toward the sun in its escape from predators; if one eye is blinded, however, its escape reaction will consist of flying in circles, which shows its dependence on bilateral optic stimulation. To illustrate that not all taxes depend on equal bilateral stimulation, this same butterfly will continue to pursue females in a straight path after unilateral blinding. The *light-compass reaction* of

Figure 7.2 **The orientation of the beetle is given by its choice of the left or right arms of the Y-maze globe it holds in its feet.** [From R. A. Hinde, *Animal Behavior: A Synthesis of Ethology and Comparative Psychology.* New York: McGraw-Hill Book Company, 1966.]

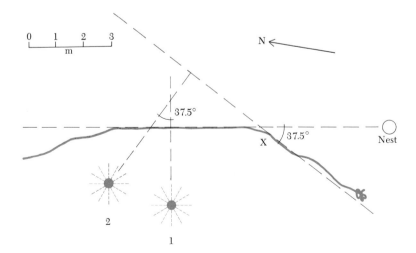

Figure 7.3 **Homing ants maintain orientation toward the nest by progressing at right angles to the sun.** When captured and held at point X while sun moves 37.5° from position 1 to position 2, their path (shown in white) continues at right angles to the sun and then deviates from the path to the nest by 37.5°. [From N. R. F. Maier and T. C. Schneirla, *Principles of Animal Psychology.* New York: McGraw-Hill Book Company, 1935, p. 136.]

ants and bees is a taxis that can also occur with only one eye. In this case, moreover, the orientation is not simply toward or away from the stimulus source, but rather the organism orients and moves at some angle to the source of light. Thus, homing ants will change their direction in accordance with the change in position of the sun even when they are captured and kept in dark boxes (Figure 7.3). Even if the sun is invisible, these insects are sensitive to the plane of polarization of light from the sky.

In the simplest case, such as that described by Loeb, a taxis may amount to a "forced" orientation or movement, in which the organism's adaptation is a simple, automatic, innate pattern of response to sensory stimulation. In other cases, however, a taxis may be part of a more complex pattern of behavior, so that a natural orientation may depend on two or more taxes, or a given taxis may be imbedded in a complex instinctive act. For example, the upright orientation of fish, ventral surface

Figure 7.4 **Reaction of Crenilabrus rostratus to light coming from different directions (arrows). In a, the fish is normal; in b, the fish has had its response to gravity eliminated by the removal of the labyrinth of the ear.** [Redrawn from N. Tinbergen, *The Study of Instinct.* London and New York: Oxford University Press (Clarendon), 1951.]

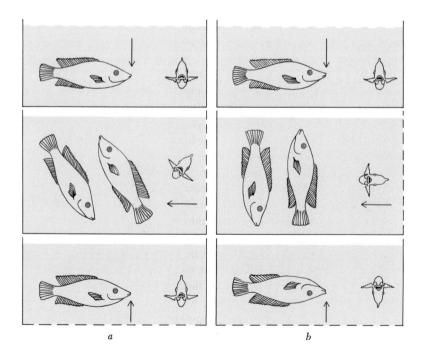

a   b

down, may depend on both photic and gravitational taxes. Thus, if light comes in from the side of a tank rather than from above, certain fish may orient at an angle upward or downward. If the effects of gravity are removed by destroying the labyrinth of the inner ear, the fish will orient perpendicularly with light from the side, and will even turn ventral side up if light comes from below (Figure 7.4).

More complication is introduced when other processes or factors *interfere* with the basic orientation. Thus, the cockroach naturally avoids light, but it can temporarily reverse this natural orientation if it is always given an electric shock when it is in the dark side of a compartment and never when it is in the lighted side. In higher animals, a simple taxis may show up under very limited conditions, but this is easily upset by the intrusion of other factors. Thus, the newborn rat, which has its eyes closed, will show a rather fixed orientation to gravity when it is placed on a wide inclined plane. Its natural tendency is to climb upward against gravity (negative geotaxis). The climb is not straight up the incline, however, but is at some angle to the left or to the right, and the size of the angle is determined by the steepness of the incline. The steeper the incline, the greater the angle of climb, approaching the straight upward path. But when the rat's eyes open, its behavior becomes more variable, and it responds visually to the edges of the inclined plane and the table top, so that it is very likely to deviate from a straight path upward and turn around and come down to the table surface. Although adult rats show some tendency to climb upward, they very quickly learn to escape from the inclined plane and will either jump off one side or turn and run down to the table.

Thus, while it is possible to demonstrate many clear-cut instances of relatively fixed and stereotyped orientations to specific stimuli in the animal kingdom, not all taxes occur so simply in nature. Very early in the study of such behavior, H. S. Jennings pointed out that all animals, even protozoa, show great variability in their behavior and are not completely fixed in their orientation responses. The reason for this is, of course, that a living organism is responsive to more than one aspect of its environment and internal state at any given instant and may be making several different adaptations at once. The higher in the phylogenetic scale we go and the more modes of adaptation an animal has, the more variable is its behavior, and thus the less fixed and stereotyped its orientation. Nevertheless, it is clear that, under appropriate conditions, lower organisms will display to external stimuli certain *relatively* fixed and stereotyped reactions that we can designate with the term taxis. And often, in observing complex patterns of behavior, we can identify component acts that are *relatively* fixed orientation responses imbedded in a more complex whole.

Very similar to the taxes are the reflexes, which are relatively stereotyped and fixed responses to stimuli that fit the definition of innate behavior in the sense that they are the outcome of inherited neural mechanisms. In fact, in many respects it is difficult to make a hard and fast distinction between taxes and reflexes. Generally speaking, taxes involve an orientation of the whole body that may involve a number of specific reflex responses. Reflexes, like the startle reflex or righting reflex, may involve all or most of the body, but, typically, they are responses of part of the body, such as the flexion of a leg in response to painful stimuli or the constriction of a pupil to intense light. Quite clearly such reflexes are adaptive and, as behavior goes, relatively invariable. Yet it doesn't take much observation of reflexes to see that there is some variation in them, especially in the higher vertebrates and in reflexes dependent on levels of the nervous system above the spinal cord.

In general, there are two classes of reflexes. The *tonic reflexes* are relatively slow, long-lasting adjustments that maintain muscular tone, posture, and equilibrium. The *phasic reflexes,* on the other hand, are rapid, short-lived adjustments such as that seen in the flexion response. Reflexes may be organized at various levels of the nervous system and may occur in varying degrees of complexity; usually those with greater complexity depend on the higher segments of the nervous system. In vertebrates, as we pointed out earlier, simple flexion and extension reflexes, including stepping, may be organized within a few segments of the spinal cord. But the coordinated alternation of flexion and extension that is locomotion is organized over many segments of the spinal cord and normally requires the influence of the midbrain. The same is true of the righting reflexes, which involve complex patterns of responses for keeping the head and body lined up and both of them upright with respect to gravity.

Many patterns of behavior are complex arrays of simple reflexes. At one time, it was believed that all complex behavior could be understood fully in these terms and that even learning and thought were nothing but complex combinations on innate and conditioned, or acquired, reflexes. Although theoretically it is possible to analyze almost any behavior into its component reflexes, this has rarely been done with any degree of success, and it has been quite clear that many kinds of behavior—instinctive patterns for one—involve something more than complex chaining of simple reflexes with invariable stimulus-response relationships.

Nevertheless, the reflex response is one of the major modes of adaptation in the animal kingdom. In the course of evolution, however, reflexes become less prominent features of behavior, for they become more variable and more and more subject to modifying influences of the higher neural mechanisms, and are overshadowed by other modes of adaptation.

By far the most complex and most fascinating of the innate behavior patterns are the instincts. Unfortunately, many misconceptions have surrounded the concept of instinct in the past, for very early in its history the term instinct implied some mysterious, vitalistic force which impelled the organism to action and directed its course with the infallible "wisdom of nature." Today we know that no special energy is released in instinctive behavior beyond the energy from the metabolic mechanisms that lie behind all the cellular activity involved in behavior. We also know that in many cases instinctive behavior is not an infallibly accurate "fixed pattern" of response, since there is too much variability in the behavior of even simple organisms, and the behavior of the higher organisms is constantly being modified and shaped by individual experience and learning. Furthermore, while it is clear that instinctive behavior is adaptive and has direction, so is all other behavior, and it is a serious teleological error to ascribe special purpose either to animals or to nature.

One early attempt to solve these difficulties was the establishment of three criteria for instinctive behavior. To be an instinct, behavior would have to be (1) unlearned, (2) characteristic of the species, and (3) adaptive. Applying these criteria proved very difficult, however. We have already pointed out that "adaptive value" is not a criterion distinguishing instincts from other behavior. The first two criteria are even more difficult, especially in the case of higher animals, for it is not easy to control individual life experiences to the point where we can be sure learning has not contributed to a pattern of behavior. For example, not all cats kill mice, although this behavior is said to be instinctive; it turns out that kittens often must see adult cats killing mice before they do so themselves, and that kittens reared with mice rarely become mouse-killers. Similarly, a chaffinch reared in isolation from its kind sings a much simpler song than chaffinches reared with adult birds (Figure 7.5), and it may never be able to learn the full song of the species if it is kept isolated from its kind past one breeding season.

Because it is so difficult to assess the criteria of instincts, many ardent believers in the concept were able to apply it all too freely in the past to almost every kind of behavior in every animal, including man. Worse than that, they used instinct as an explanatory concept, never seeking

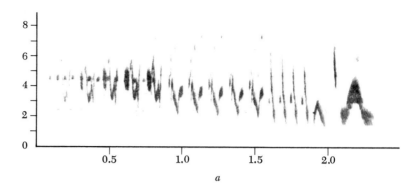

a

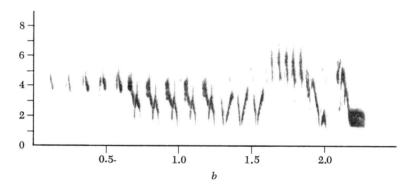

b

Figure 7.5 **Records of the patterns of sound frequencies in normal chaffinches (a and b) and those reared in isolation from adults of their species (c). The isolated birds have only an incomplete song, showing the role of experience in the development of instincts.** [From W. H. Thorpe, *Learning and Instinct in Animals.* Cambridge, Mass.: Harvard University Press, 1958.]

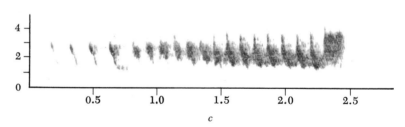

c

to analyze or investigate its underlying mechanisms. Thus it was said, for example, that man fought because he had a fighting instinct and that that was all there was to it.

Modern understanding of instinctive behavior came about through the abandonment of these misconceptions in favor of direct, experimental investigation of these complex behavior patterns. Today, two rather different scientific schools of thought are converging in their experimental investigations. One is that of the European ethologists, who are zoologists and who investigate behavior under natural or seminatural conditions.

In their approach, they have sought full knowledge of the behavioral repertory of the species they study in order to see instinctive behavior in its proper context. Their interest has been chiefly in animals other than mammals, and they have focused mainly on parental and filial behavior, social behavior, and reproductive behavior. The second group is made up primarily of American psychologists and physiologists, who are interested mainly in hunger, specific hunger, thirst, sexual behavior, temperature regulation, sleep, rage, and fear. These investigators have worked primarily with mammals, including primates and man himself. Typically, they have been interested in one limited aspect of behavior, observed under artificial laboratory conditions, but have probed deeply for its neurophysiological mechanisms. Despite the differences in these approaches of these two scientific schools, it is remarkable how similar their findings and basic conceptions are.

We can begin with the viewpoint of the ethologists, for they have offered the most complete and most general conceptions. They make two important points that distinguish instinctive behavior from taxes and reflexes. First they stress how instinctive behavior so often depends on some special condition of the internal environment of the organism. For example, many aspects of reproductive behavior depend on the presence of the sex hormones, with the result that at one extreme there is no positive response to strong sexual stimulation in the absence of hormones, and at the other, when hormone concentration is high, only minimal stimulation is required to elicit a complete pattern of sexual behavior. In some cases, the internal state may be so strong as to lead almost directly to behavior without any measurable eliciting stimuli; this is the so-called *vacuum reaction*. The second point is that stimuli serve only to trigger instinctive behavior and are not always necessary to guide it through the entire pattern. The gray goose, for instance, will retrieve an egg that has rolled outside its nest by pushing it between its legs with the underside of its beak, and it may continue such diligent pushing movements to "completion" even if the egg has rolled away from it.

The ethologist conceives of instinctive behavior as a complex interaction of both internal and external influences, organized in a hierarchy of neural mechanisms, with each level of the nervous system controlling specific instinctive acts. For example, the reproductive behavior of the male three-spined stickleback fish is made up of migratory, territorial, fighting, nesting, mating, and parental instinctive acts. Migration, believed to be organized at the highest neural level, is set off by increases in gonadal hormones produced by seasonal increases in daily light. Directed by temperature, the fishes move to warm, shallow fresh water, where they select a territory in response to the sight of green vegetation. Here they build a nest, defend the territory against intruding males, and

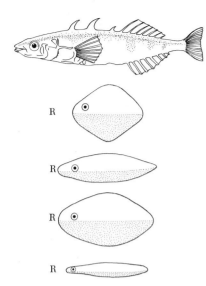

Figure 7.6 **Fighting in the stickle-back was elicited by the lower 4 models, which did not have the shape of the fish, but did have the red underside. The faithful model (top), which lacked the red belly, did not elicit fighting.** [Redrawn from N. Tinbergen, *The Study of Instinct.* London and New York: Oxford University Press (Clarendon), 1951.]

attract females by the increased redness of their bellies and by executing a zig-zag dance, which leads the female to the nest. Finally, mating takes place, and when the eggs are fertilized, the males participate in their care by fanning them with movements of their fins.

Each of the instinctive acts, is triggered by the combined effects of external stimuli, hormones, and excitatory central neural influences. Investigation of the external stimuli usually shows them to be complex spatial and temporal patterns, which can be analyzed into specific components called "sign stimuli." These may be physical aspects of the environment, such as light, water temperature, vegetation, nesting materials, and so on. Or they may be some aspect of a living animal, such as the swollen abdomen of the gravid female, the red belly of the intruding male, or the male's assumption of a threatening posture with nose downward. When the sign stimuli are produced by the behavior of another animal, we have an innate basis for social interactions and organization.

Careful experiments by the ethologists have shown how complex, and often how specific, these sign stimuli may be. Thus, territorial defense in the stickleback may be elicited by a dummy fish with a red belly even if the dummy is not shaped like a stickleback. On the other hand, a faithful model of the stickleback fails to release fighting if it does not have the red belly (Figure 7.6). Furthermore, it has been shown that more vigorous attack will be made if the stickleback is presented in the vertical position with nose downward rather than horizontally (Figure 7.7).

Similar analyses have been made of other instinctive acts in other animals. The nestlings of the herring gull, for example, beg for food when presented with dummy gull heads. However, the strength of this response depends on, among other things, the presence of a spot on the model's beak and on its color (Figure 7.8). In addition the direction of the begging response is determined partly by the red spot and partly by the tip of the beak (Figure 7.9). The innate response, therefore, is determined by the whole pattern of stimulation provided. The importance of the configuration is seen even more dramatically in the escape reactions of young ducks and geese in response to the short neck of a bird of prey. When the dummy shown in Figure 7.10 was sailed to the right, with the short neck leading, the escape reactions were elicited as though in response to a bird of prey. But when the dummy was sailed to the left, the long neck leading, presenting the pattern of a goose, there was no sign of disturbance in the young birds. This experiment, of course, says nothing about whether the "escape reactions" were innate or learned, and while recent work shows that birds adapt to both directions of movement and respond to neither after repeated

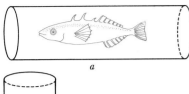

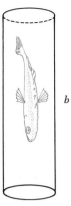

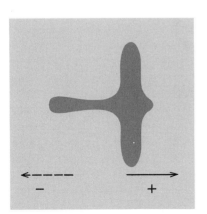

Figure 7.7 *When the three-spined stickleback is prevented from assuming its natural "threat-posture" (a), it elicits far less fighting from other fish than the fish in b which is placed in the "threatening" posture.* [Redrawn from N. Tinbergen, *The Study of Instinct.* London and New York: Oxford University Press (Clarendon), 1951.]

Figure 7.8 *Experiment with herring gull models shows that the strength of the begging response of the young (length of bar to right) depended on a spot on the beak, especially on a dark spot and on a red one.* [Redrawn from N. Tinbergen, *The Study of Instinct.* London and New York: Oxford University Press (Clarendon), 1951.]

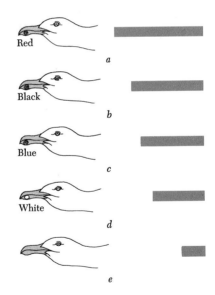

Figure 7.9 *The begging response of the herring gull is directed about as often to the mouth (70 times) as it is to the red spot (74 times), which was shifted in this experiment from the beak to the head of the model.*

Figure 7.10 *When the model was sailed to the right, it elicited flight and escape in young ducks and geese; sailed to the left, it had no effect.* [Redrawn from N. Tinbergen, *The Study of Instinct.* London and New York: Oxford University Press (Clarendon), 1951.]

trials, the original observations of a differential effect of the "hawk" and the "goose" pattern in initial trials is most instructive.

Although the innate aspects of instinctive responses have been emphasized here, it is clear that they may often be modified by experience and learning. Indeed, in some cases, as in imprinting (page 109), experience determines the specific sign stimulus that will be effective. Even the release of hormones, such as those crucial to the reproductive behavior of the ringdove, may depend upon prior experience with appropriate sensory stimulation (see page 136). So hormones and sign stimuli, acting together with past experience, determine the expression of instinctive patterns of behavior. As we shall see in a moment, the same conclusion is reached by the American students of motivated behavior, working mainly with the mammal.

MOTIVATED BEHAVIOR

The approach of American workers to problems of this sort is somewhat different. On the behavioral side, many American psychologists emphasize the *motivational* aspects of instinct and start with the conception that many patterns of instinctive behavior can be analyzed into *drive* directed toward a *goal,* the attainment of which results in reduction of the drive, or *satiation.* These terms may be illustrated by the case of a three-year-old boy who had an abnormal craving for salt. From early life, he always preferred salty foods and would lick the salt off bacon and crackers rather than eat them. When he was 18 months old, he discovered the salt shaker and began eating salt by the spoonful. He learned very quickly to point to the cupboard and scream until he was given the salt shaker, and the first word he learned was "salt." It turned out that his craving for salt had kept him alive, for when he was taken to the hospital for observation and placed on a standard hospital diet with limited salt, he died within seven days. At autopsy, it was learned that he had tumors of the adrenal gland and thus lacked the hormones necessary to reabsorb salt through the kidneys. Only by constantly replacing salt lost in his urine could he maintain himself.

From this example, it can be seen that *drive* is a striving toward some goal (salt in our illustration). It is reflected in increased activity, a willingness to work or to overcome some resistance in order to achieve the goal, and often by learning new instrumentalities toward achieving the goal (screaming, pointing, saying the word "salt"). The *goal* itself may be an object that is acted upon or ingested, as in the case of salt, or it may be the execution of a pattern of behavior, as in mating. Or still more generally, it may be a change in the stimulation of the animal, as in the escape from painful stimulation. New goals may be learned if they are instrumental in attaining the natural goal. Thus, a chimpanzee learns to work for poker chips, which he later can redeem for food and water. *Satiation* is drive reduction. It is characterized by a reduction in activity and in willingness to work for the goal. *Motivated behavior,* then, is a drive that leads to goal-direct behavior and satiation. It may be measured by the intensity or rate of *consummatory behavior,* such as in eating, drinking, and mating, or by the rate or intensity of work the animal will do to reach either the goal itself, some small fraction of it, or a learned goal such as the poker chips the chimpanzees associate with food.

On the neurophysiological side, motivated behavior, or instinctive behavior, has been investigated by tracing out the parts of the nervous system involved and the effects produced on them by external stimuli and changes in the internal environment. The focus of attention in work on mammals has been the *hypothalamus* because it has been found to contain *excitatory mechanisms,* whose actions contribute to the arousal of motivated behavior, and *inhibitory mechanisms,* whose actions contribute to the reduction of motivated behavior. For example, in the case of feeding, investigators have discovered that destruction of the ventromedial regions of the hypothalamus on both sides results in a vast increase in eating, to the point where a rat or cat or monkey might double or triple its body weight. The ventromedial area qualifies, therefore, as a part of the inhibitory mechanism. Since bilateral destruction of the lateral regions of the hypothalamus will cause an animal to starve to death, even in the presence of its normal food, these regions can be thought of as containing an excitatory mechanism. These ideas are borne out when electrodes are chronically implanted into these regions of the brain and just their tips are electrically activated while the animal is awake and active. In such a prepared animal, electrical stimulation of the lateral hypothalamus produces increased eating, and stimulation of the ventromedial hypothalamus causes a reduction in eating.

Similar, rather specific hypothalamic mechanisms have been found for thirst, sexual behavior, emotional behavior, sleep, and maternal behavior, and these findings have led to the notion that drive is based on the activity of an excitatory hypothalamic mechanism and that satiation is based on the activity of an inhibitory hypothalamic mechanism. It is also believed that the action of both these mechanisms is controlled by relevant sensory stimulation, changes in the internal environment, and influences from the cerebral cortex (Figure 7.11). For example, it has been shown by recording the electrical activity of the hypothalamus that genital stimulation results in the rather specific activation of a small region of the hypothalamus whose destruction leads to the abolishment of sexual arousal. Experiments showing that appetite-depressant drugs selectively activate the inhibitory feeding mechanism in the ventromedial

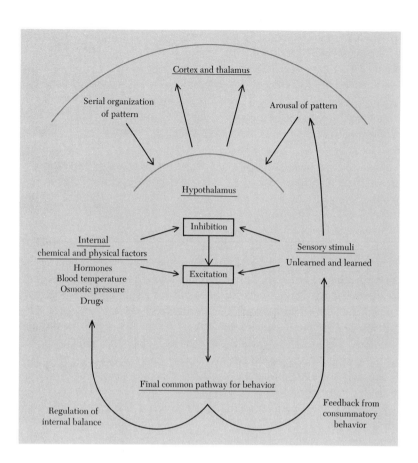

Figure 7.11 **Schematic diagram of the physiological factors controlling the excitatory and inhibitory hypothalamic mechanisms that govern motivated behavior.**

hypothalamus illustrate the influence of the internal environment. Even more dramatic are the cases in which small pipettes are implanted chronically into the region of the hypothalamus implicated in sexual behavior. Here injection of minute quantities of sex hormones produce prompt and vigorous mating behavior.

This type of experiment suggests there may be special chemical receptors in the hypothalamus that are selectively sensitive to the level of circulating sex hormones. Other experiments suggest that the hypothalamus may also contain temperature receptors and osmoreceptors. For example, injection of a few thousands of a cubic centimeter of hypertonic saline into the hypothalamus of a goat will lead to prompt drinking of as much as 7 liters of water, indicating that osmoreceptors may be involved in registering the cellular dehydration of thirst. It may be through mechanisms of this sort, then, that the internal environment plays its role in instinct and motivated behavior.

The role of the hypothalamus and other rhinencephalic structures in motivation can be seen on an even more general basis if a somewhat different kind of experiment is performed. In this case, an animal such as a rat, cat, or monkey, is prepared with an electrode chronically implanted in the lateral hypothalamus, the septum, or the tegmentum. The animal is then placed in a situation in which every time it pushes on a lever, a brief current is turned on, stimulating a point within the brain. Very quickly the animal comes to push the lever vigorously and rapidly, stimulating its own brain many times a minute for hours on end. That such self-stimulation behavior is strongly motivated is shown by the fact that the animals will push the lever many times for just a single stimulation, and, even more dramatically, they will walk across an electrified grid to push the lever even more readily than they will cross the grid to eat when they are very hungry. Some places within the brain appear to be negative rather than positive, in that animals will stop pushing the lever if they are stimulated, or they will work hard to end such stimulation after the experimenter initiates it. In general, it turns out that negative self-stimulation points lie along the midline of the brain, while positive self-stimulation points lie more laterally in the brain.

The role of the cerebral cortex is perhaps most clearly seen in emotional behavior. Here it has been found that destruction of the neocortex produces extreme placidity in cats, which suggests that the neocortex is essential for the arousal of aggresive behavior. On the other hand, destruction of parts of the old cortex results in unbridled ferocity. These cortical mechanisms, it is believed, have their influence through hypothalamic mechanisms known to be important in the arousal and control of aggressive behavior.

It is not surprising that, in the course of investigating the controlling neurophysiological mechanisms in many different species, we can trace out rather orderly changes in phylogeny. For example, in sexual behavior, the relative contributions of hormones, sensory stimuli, cortical influences, and learning undergo systematic changes in the mammalian series. If we compare rodents, carnivores, and primates, including man, we find a decreasing dependence on sex hormones and an increasing dependence on sensory stimuli, the cortex, and learning, to the point where the higher primates and man may not require hormones, but rather may be crucially dependent on learning. For example, whereas a rat reared in isolation will mate successfully in its first experiences, a monkey isolated past puberty may be totally incapable of mating behavior.

In feeding behavior, similar changes can be traced over even a wider range of the animal kingdom. The blowfly responds with chemoreceptors to the chemical aspects of food and, through the influence of its central nervous system, either accepts or rejects foods. The stronger the stimulus, the stronger the reaction regardless of the actual food value of the chemical. Eating proceeds until sensory adaptation occurs or until an internal stimulus associated with the presence of food in the foregut is produced. If the nerves from the foregut are severed, there is no lasting cessation of eating. The blowfly, therefore, is quite stimulus-bound.

The rat is also highly responsive to the stimulus aspects of food and will eat nonnutritive sweet substances, such as saccharin. At the same time, however, it may regulate its intake according to the caloric content of food, and it is strongly affected by previous experience or learning. For example, a thiamine-deficient rat selects food containing thiamine. But if this food is first flavored with anise and the anise is then shifted to a food without thiamine, the deficient rat will forego the thiamine that will fulfill its need and select the anise-flavored, non-thiamine food in accordance with its habit.

Even more dramatic is the case of adrenalectomized rats. These animals, like the boy with the adrenal tumors, typically make up for the salt they lose in the urine by drinking large quantities of strong salt solutions, and thus manage to survive. If they have had the opportunity to drink sugar water before being operated on, however, they continue to prefer it to salt solution and die of salt deficiency. Controls that never had sugar solutions before operation and thus never developed a habit of sugar preference show a preference for salt over sugar and survive adrenalectomy. Habit, therefore, can be strong enough to be fatal. It plays an even greater role in man than it does in animals, for the regulation of our food intake and food preferences and aversions are as much a matter of early life experiences, personality, dietary laws, and social culture as they are of nutritional requirements.

SUMMARY

In this chapter, we have discussed a variety of stereotyped behavior that serves in the adaptation of an organism, from simple taxes and reflexes through complex instincts. Taxes are innate responses of the whole organism in orientation toward or away from stimuli; they are the result of inherited properties of the organism's receptors and central neural connections and are prominent in the adjustments of the lower invertebrates to their environment. Reflexes are also innate sensory-motor responses, usually of a part of the body. These are seen in all metazoa and are likewise determined by receptor properties and central neural connections. But reflexes may complexly interrelate with one another, sometimes in elaborate patterns of response, and with the development of learning capacity, they may readily be modified, as we shall see in Chapter 8.

Instincts are the most complex of the stereotyped behaviors. In the lower organisms, they are innate responses that are elicited by the combined influences of the internal environment and sensory stimulation. In a sense, the internal environment "primes" the response mechanism, and the sensory stimulation triggers it, with the result that a complex neural mechanism may be set into action, yielding a complex sequence of behavior. In the higher organisms, this innate mechanism may be greatly modified by learning, and indeed overshadowed by learning, to the point where it is difficult to recognize instincts as such. By this time in phylogeny, the motivational aspects of instinctive behavior emerge clearly, and it is quite meaningful to talk about "drives," "goal-directed behavior," and "satiation." In mammals, it is believed that the arousal of drive is the result of the activation of an excitatory neural mechanism in the hypothalamus and that the reduction of drive, or satiation, is the result of the activation of an inhibitory hypothalamic mechanism. It is through these basic mechanisms that the combined effects of the internal environment, sensory stimuli, cortical influences, and learning regulate motivated behavior.

OF ALL THE BEHAVIORAL CHARACTERISTICS OF LIVING ORGANISMS, PERHAPS none is as striking as the ability to learn. This is the process through which life experiences leave their mark on an individual and the one that permits an animal to develop new adaptations in the light of past experiences, and sometimes to develop what turn out to be maladaptations. There are many kinds of learning, ranging from the simplest modifications of innate behavior to the most complex, symbolic transactions seen in the reasoning of man. All, however, are characterized by an enduring change in the behavior of the organism, perhaps a permanent change. From a biological point of view, the change in behavior must be a change in the functioning of the nervous system, and, if it is a permanent change, perhaps it is also a change in the structure of the nervous system.

By insisting that the change in learning must be a lasting one, we automatically rule out the transient changes resulting from sensory adaptation, fatigue, and fluctuations in motivation. Changes due to growth and maturation are harder to distinguish, for they are also permanent ones, and often it takes careful experimental procedures to separate the effects of maturation from the effects of learning. One obvious way is to try to hold experience constant over the period of time maturation can be expected to occur. This has been done successfully in the case of swimming behavior in salamanders, where one group of animals was anesthetized just before swimming movements developed and was not released from anesthesia until after a control group of animals had developed to the point where they were swimming normally. Since the experimental salamanders swam essentially normally as soon as they recovered from the effects of anesthesia, it is obvious that this change in behavior was mainly a matter of maturation rather than the result of experience or learning.

Experiments such as these are, of course, much harder to do in the higher animals, but an approximation has been successfully achieved in man. In this case, identical twins were used. One twin was allowed extensive experience in climbing stairs and the other was restricted to flat surfaces. This time, maturation was held relatively constant and experience allowed to vary. Again, experience turned out to be a relatively insignificant factor, for the second twin was as good as the first when finally given the opportunity to climb stairs.

Despite the value of these experiments, the most direct and obvious way to be sure that learning is taking place is to associate the change in behavior with some deliberate training procedure. This is the method employed by students of learning who are interested in investigating the phenomenon itself and in specifying the natural laws governing the process. This is the method, in its various forms, that we shall mostly discuss in this chapter. Our purposes will be: first, to describe various kinds of learning, from simple to complex; second, to try to arrive at general laws of learning; third, to see how the capacity to learn varies in phylogeny; and fourth, to summarize what we know of the neurophysiological basis of learning.

We can begin our discussion with a highly specialized and limited form of learning called *imprinting*. This is a phenomenon seen most clearly in birds during the very early period of their lives following hatching. In a young bird it consists simply of learning to follow the first large, moving object it sees and hears in a manner reminiscent of the natural tendency of the bird to follow its mother. For example, if a duckling is hatched in the presence of a large green box containing a ticking alarm clock, it will follow the movement of the box along a trolley wire. After some exposure to the box, the duckling will even follow it rather than following its own mother or other birds. Quite clearly this is a case of learning, for the bird can be imprinted in this manner on almost any object, animal, or person. But it must depend on some special condition of the nervous system prevailing only early in development, for if the bird is not imprinted soon after hatching, it may never imprint at all.

Perhaps the simplest kind of learning seen throughout life is *habituation*. In this type, upon repeated exposure to a stimulus, an animal gradually decreases its natural response until it may disappear entirely. Thus, an animal will orient toward a moderate sound, but with each successive exposure, it orients less and less until it orients no longer. That this change is not merely fatigue or sensory adaptation is clear, for the decrement in response grows with daily exposures to the stimulus and will last over long periods of time without stimulation. In a sense, habituation represents the dropping out of responses that are of no "significance" in the life of the animal. Most of the rest of learning is concerned with the strengthening of responses that *are* of significance so that they may be evoked more readily and with increased frequency or probability.

One very simple form of learning of the latter type is *classical conditioning,* so called because it was discovered by I. P. Pavlov, the father of conditioning. In his classical experiment, Pavlov lightly restrained a dog in a harness, repeatedly blew meat powder into its mouth, and recorded accurately the amount it salivated. Then he associated the sound of a bell with the meat powder, repeating this procedure a great many times at successive intervals. The bell did not at first elicit salivation, of course, but after repeated pairings with meat, it came to do so. In describing this experiment, Pavlov called the salivation to the bell a *conditioned reflex* (CR), the bell a *conditioned stimulus* (CS), the salivation to the meat an *unconditioned reflex* (UCR), and the meat itself an *unconditioned stimulus* (UCS). The same experiment has been repeated many times on different animals and with many different stimuli and responses. For example, the UCR may be a flexion of a leg in response to electric shock to the foot (UCS), and if this reflex is paired with the sound of a metronome (CS), that signal will eventually cause a leg flexion (CR) (Figure 8.1). Typically, the CR is similar to the UCR, but it is never completely identical to it. Thus, the best way to describe classical conditioning is as *a process in which a previously neutral stimulus (CS = bell) is enabled to elicit a response (CR = salivation) that it never elicited before training.*

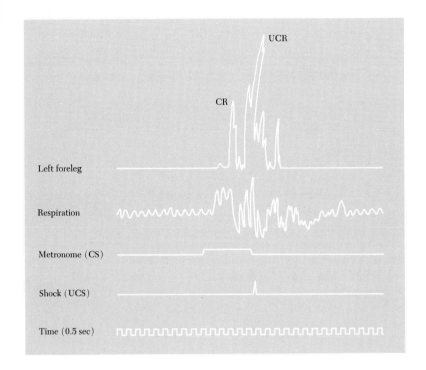

Left foreleg

Respiration

Metronome (CS)

Shock (UCS)

Time (0.5 sec)

UCR

CR

Figure 8.1 **Graphic record of a conditioned leg flexion and a conditioned respiratory response in the sheep, following the pairing of the sound of a metronome with shock to the foot for many trials. Note that the unconditioned flexion response is larger than the condition response, and also that there is a conditioned respiratory response that precedes the conditioned flexion.** [Redrawn from E. R. Hilgard and D. G. Marquis, *Conditioning and Learning.* New York: Appleton-Century-Crofts, 1940.]

In his experiments, Pavlov found that the time relation between the CS and the UCS was critical. If the UCS preceded the CS, there was very little, if any, conditioning, and if the CS preceded the UCS by more than about a second, conditioning became more and more difficult to establish (Figure 8.2). Furthermore, he found that as he repeatedly paired the CS and UCS, the strength of the CR grew steadily throughout the course of *acquisition* of the CR (Figure 8.3). On the other hand, if he presented the CS alone repeatedly, the strength of the CR decreased steadily in what he called the *extinction* of a conditiond response. He concluded, therefore, that there was something about the presentation of the UCS that was essential for the strengthening and maintenance of the conditioned response. This "something" he called *reinforcement.* Some workers believe that reinforcement is a reward involving the reduction of a drive, because the dog received meat powder in one case and escaped from painful stimulation of the foot in the other. We shall shortly see that in some cases reinforcement appears quite clearly to be a drive reduction or reward. In classical conditioning, however, Pavlov believed that the reinforcement was nothing more than some effect in the nervous system produced by eliciting the uncondi-

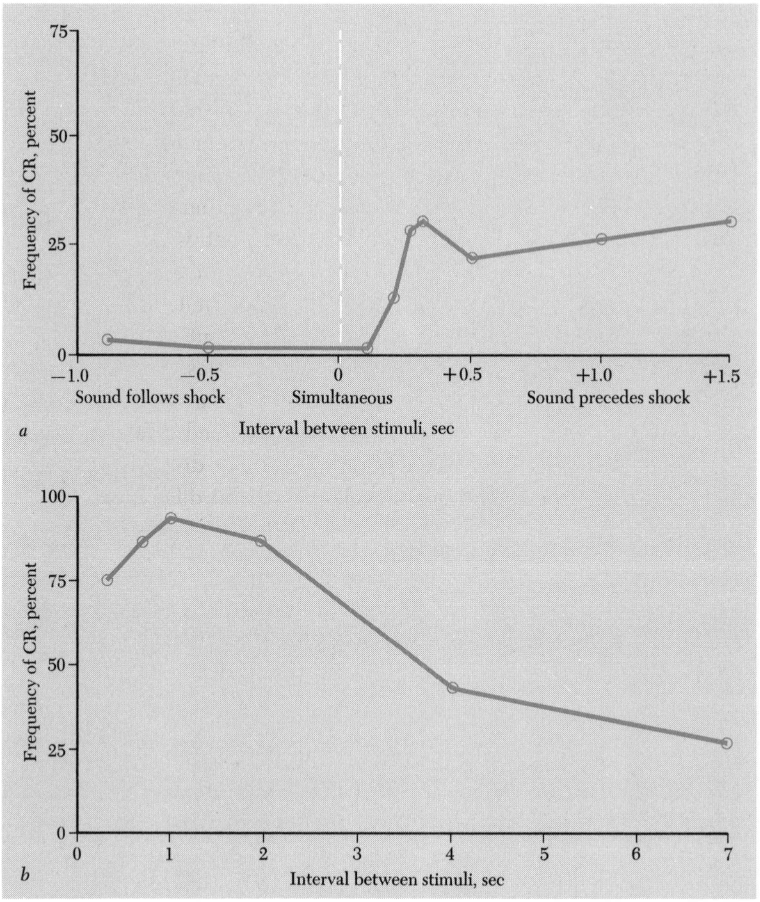

Figure 8.2 Two separate experiments illustrate the importance of temporal relations between the CS and UCS in conditioning: (a) the difficulty of establishing "backward conditioning" when the UCS precedes the CS in human eyelid conditioning; (b) the reduction in the strength of conditioning as the CS-UCS interval increases in conditioned respiratory response in the rat. [Redrawn from E. R. Hilgard and D. G. Marquis, Conditioning and Learning. New York: Appleton-Century-Crofts, 1940.]

tioned response during presentation of an ongoing conditioned stimulus.

Extinction, Pavlov believed, is a new learning that results in an inhibition of, or interference with, the CR. The reasons for this belief are as follows. If, early in training, a distracting stimulus (sudden noise) occurs as the CS is presented, there will be a great reduction in the CR; this is called *external inhibition*. If the same distracting stimulus occurs early in extinction, there is an increase in the strength of the CR, called *disinhibition* because presumably it is the result of removing the inhibiting or interfering with effects of new learning by external inhibition of that new learning.

Two other concepts of Pavlov's will round out our summary of classical conditioning. One is the concept of *generalization*, which derives from the fact that an animal conditioned to one stimulus—say a 1,000-cycles-per-second (cps) tone—will also be conditioned to some degree to other, similar stimuli—say a 500- or a 1,500-cps tone. Actually, there is a gradient of generalization in the sense that the farther away a stimulus is from the original CS along some continuum, the weaker is its capacity to elicit a CR. The second concept is *discrimination,* which is the result of reinforcing the CS (1,000-cps tone) and extinguishing a similar stimulus (say 1,500-cps tone) to the point where the animal always responds to the 1,000-cps tone and never to the 1,500-cps tone. As Pavlov pointed out, these conditioning and discrimination techniques not only afford a way to study animal learning, but also provide the opportunity for objective study of sensory capacities. For example, if we wish to learn something about the sensitivity of a dog's hearing, we can use the discrimination-training method described above to tell us the smallest difference in frequency the animal can distinguish, a value which turns out to be as low as a 2-cycle-per-second difference.

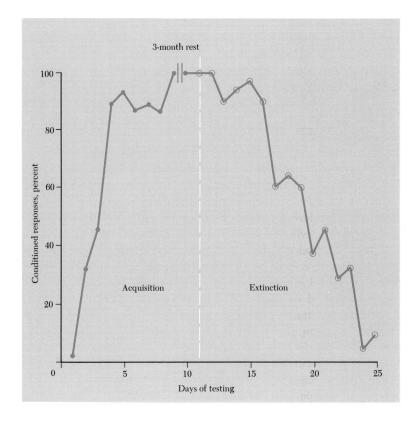

Figure 8.3 *The acquisition and extinction of a conditioned leg flexion response in a cat where a tactile CS was used, showing the increasing strength of the CR as a function of the number of trials over which reinforcement is given (acquisition) and the decreasing strength of the CR as a function of the number of trials without shock reinforcement (extinction). Note, however, that there is no decrement in CR strength over a 3-month rest period.*

From these facts about classical conditioning, we can derive a number of general laws that are remarkably similar to the laws of association described by the English association philosophers and later by students of human verbal learning. The first is the *law of contiguity,* which says that items to be associated must occur together in time and place. We have seen how conditioning grows less and less effective as the interval between CS and UCS-UCR is lengthened. The second is the *law of repetition:* Pavlov's studies showed that the strength of a CR grows progressively as more and more pairings of CS and UCS are made. The third law is the *law of reinforcement,* which describes a process that is essential to the strengthening of a CR. In certain forms, this law may be identical to the *law of effect,* which states that reward strengthens associations and punishment weakens them, presumably by training interfering escape and avoidance responses. The fourth law might be called the *law of interference;* it covers the case of extinction, or forgetting, and states that a conditioned response may be weakened and indeed inhibited by new learning that interferes with it.

The case for an interference concept of extinction and forgetting is perhaps best argued on the basis of data on human verbal learning. Here the degree of forgetting is a function of the activity intervening between learning and test of memory. If the intervening activity involves learning of verbal material similar to that originally learned, interference is marked and forgetting will be greatest. If the intervening activity is totally different from the material originally learned, then forgetting is slight. And most interesting of all, if a person goes to sleep after the original learning, there is the least forgetting in the test of memory.

If extinction and forgetting are the result of interference, then it may be that *learning is permanent* and is merely covered over or inhibited by new learning. Pavlov's discovery of disinhibition supports this notion, and so do the studies where learning survives so completely through a period of sleep. Also, the maintenance of the strength of the CR over a 3-month rest period shown in Figure 8.3 lends strength to this idea. But even more convincing are experiments in which pigeons were trained in a simple response and then removed from the learning situation so that there could be no possibility for new learning that would interfere, that is, no opportunity for extinction. After 10 yr, these pigeons retained their original learning about as well as after a few days.

The case of classical conditioning is one of passive learning during which the experimenter elicits the desired response by reflex and presents the CS on schedule. The animal can do very little about it. He has to salivate or flex his leg each time the CS is paired with the UCS, and his response produces no change in this situation. The experimenter has virtually complete stimulus control over the animal and almost completely controls his responses, and the learning is distilled down to the point where the animal simply learns to make a new response to the stimulus.

Many kinds of learning are more complicated than that seen in classical conditioning since animals have some control over the stimuli they receive and often the responses they use, and their behavior has some effect on their situation. For example, an animal may learn to depress a lever or a swtich in order to get food delivered to him, to escape from electric shock, or simply to escape from confinement (Figure 8.4). In these cases, the animal's behavior is instrumental in bringing about some significant change in his environment. Thus, we call simple instances of this kind of learning *instrumental conditioning*. Although the animal is now in a quite different training situation, the same phenomena of acquisition and extinction, and the same laws of contiguity, repetition, reinforcement, and interference, apply. In instrumental conditioning, the animal starts out naïvely emitting a variety of responses that are in his natural repertory. The trainer may then select one response to reinforce: depressing a lever, standing on hind legs, turning right rather than left at the end of an alley, and so on. Because of the reinforcement, this response is emitted with greater frequency, and other responses drop out, or habituate.

Where the reinforcement involves the satiation of a drive, as in the case of the hungry animal depressing a lever or turning right to receive food, we call the training *reward training*. Where the reinforcement involves the escape from some noxious situation, such as electric shock, bright lights, or cold water, it is *escape training*. And where the reinforcement involves the possibility of avoiding noxious stimulation

Figure 8.4 **A cat operating a bar to unlock a door and escape.** [Redrawn from N. L. Munn, *The Evolution and Growth of Human Behavior.* Boston: Houghton Mifflin Company, 1955.]

altogether, we call it *avoidance training*. For example, an animal receives a shock 5 sec after a tone begins. At first, he can only escape the electric shock after it comes on by depressing a lever or leaping over a hurdle into a "safe" compartment, but later he may make the response within the 5-sec period between the onset of the tone and the onset of the shock and thus may avoid the shock entirely. According to Pavlovian principles, one would expect the CR in this experiment to start extinguishing once the animal begins avoiding the shock (UCS) consistently, but extinction proves very difficult and the animal may continue responding for many, many trials without receiving shock. Apparently, the reinforcement in this situation is more than the shock itself. It has been suggested that this may be a special case of emotional reinforcement involving a "fear" of the shock that is reduced each time the animal makes the avoidance response. But this answer is by no means certain.

TRIAL-AND-ERROR LEARNING

Instrumental conditioning can be made complicated simply by increasing the complexity of either the stimulus situation or the response possibilities. This can be done by giving an animal a choice of stimuli to respond to, one with one response and the other with another response. For example, the animal may be confronted with two doors, one light and one dark. Food is available behind the light door; the dark door is locked, and the animal must learn to approach or jump to the light door whether it appears to the right or to the left (Figure 8.5). The animal here is required to make a discrimination of brightness, but the

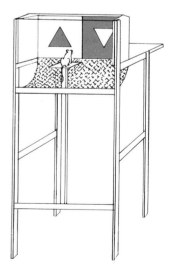

Figure 8.5 **A rat faced with a black-white discrimination in the Lashley jumping stand.** *If the black circle is correct, its door is not locked and food will be found behind it regardless of whether it is on the left or on the right. The other door, with the white circle, is always locked.* [Redrawn from N. L. Munn, *The Evolution and Growth of Human Behavior.* Boston: Houghton Mifflin Company, 1955.]

same test could be run using two different colors, a triangle and a circle, a loud and soft tone, and so on. In this type of situation, the animal is more obviously involved in a process of *trial-and-error learning,* in which correct responses are encouraged by reward and incorrect responses discouraged by withholding reward or, in some cases, by administering punishment.

Still more complicated is the multiple-choice maze in which the animal may be required to make a string of discriminations in order to thread its way through the maze to the food at the end (Figure 8.6). In a typical maze, there may be several sensory cues at each choice point that the animal could use as a basis for discrimination. Different animals may use different cues at the same choice point; we can tell whether one rat uses a visual cue and another rat an auditory cue at a particular choice point because putting out the light causes only the first rat to make an error and eliminating auditory cues will affect only the second rat. The maze, then, is under multiple sensory control, and an animal may be left free to use any one of a number of alternative cues at each choice point.

We have gone now from the simplest stimulus-response training of classical conditioning through more and more complicated instances of instrumental conditioning in which the sensory stimuli involved become more numerous and more complex and in which the animal is given a greater and greater measure of choice in the stimuli it uses. In general, as we go from the simple to the complex learning tasks, learning becomes more difficult and more easily disrupted by brain injury. With these points in mind, it will be interesting to compare the learning ability of different animals that represent different levels of the phylogenetic scale and thus different levels of development of the nervous system.

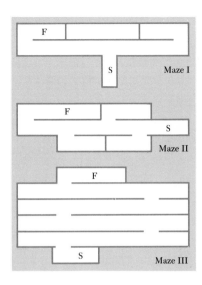

Figure 8.6 **Three mazes of graded difficulty used by Lashley to study the effects of brain lesions on learning in the rat: S is the start box and F the food box.** [Redrawn from K. S. Lashley, *Brain Mechanisms and Intelligence.* Chicago: University of Chicago Press, 1929.]

The first question we may ask in comparing the learning ability of different animals is where on the phylogenetic scale the capacity to learn emerges. Is it a basic property of all animals? Or does it depend on development of the nervous system? If so, what properties of the system are required?

Thus far, all efforts to demonstrate learning in protozoa (page 8) have failed to yield conclusive positive results. Simple metazoa, the micrometazoa, are capable of habituation, but these modifications in behavior last only a matter of minutes and are thus indicative of only minimal learning ability. Laboratory efforts to establish learning in echinoderms have consistently met with difficulties, but now recent studies of the starfish under more natural conditions have successfully demonstrated associative learning (page 29). Quite clearly, the degree

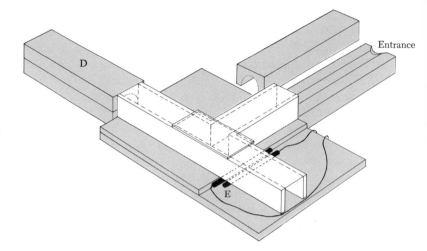

Figure 8.7 A T-maze used by Yerkes for training earthworms to turn to the right to enter a dark, moist chamber (D) and to avoid electric shock at E on the left. [Redrawn from N. R. F. Maier and T. C. Schneirla, *Principles of Animal Psychology.* New York: McGraw-Hill Book Company, 1935.]

of success obtained in a learning experiment with the simpler invertebrates depends a great deal upon the methods and techniques used. We must be careful not to conclude, therefore, that learning in protozoa is impossible or that the simpler metazoa are incapable of associative learning, for we have by no means been able to test their limits.

On the other hand, we can conclude with confidence that there is unequivocal evidence for associative learning at the level of worms in whom a bilaterally symmetrical, synaptic nervous system is already developed. Here we see clear instances of habituation and associative learning. In one experiment, earthworms were trained to go to one arm of a T maze leading to a dark, moist chamber and to avoid the other arm, which led to an electric shock and an irritating salt solution (Figure 8.7). In this simple learning of a position habit, an average of about 200 trials were required to reach a criterion of 90 percent correct responses. Interestingly enough, the worms were able to retain what they had learned after removal of the first five body segments, containing the cephalic ganglion; untrained worms could learn with the head ganglion removed. Apparently, the neural changes involved in learning can occur in the ganglia of the lower body segments.

Similar evidence has been offered in the case of the flatworm (planarian). In this case, a classical conditioning technique was used. When the worms were gliding along in a trough of water, a light was turned on, followed 2 secs later by an electric shock that caused the worms to contract longitudinally. After 150 trials, the worms contracted to the light alone over 90 percent of the time. Following this, the worms were cut in half and allowed 4 weeks for regeneration. Both the regenerated head and tail sections showed a high degree of retention of what had been learned earlier. Even after these regenerated worms were cut and a second period of regeneration occurred, there was retention of the conditioned response. Apparently, in this case, too, learning is not confined to the anterior portion of the nervous system, but has an effect throughout its extent.

A word of caution is necessary here, for many factors other than learning can increase the planarian's responsiveness to light. Worms that are kept under suboptimal conditions become hyperirritable to light. Too much exposure to intense light can also increase irritability. Even more disconcerting, some experiments suggest that electric shocks alone may "sensitize" flatworms so that they respond to the light alone even though the light and shock have never been presented together in training. Furthermore, the regeneration experiments must be done most carefully because planaria are very sensitive to light in the early stages of regeneration before they have attained full size, and it is easy to mistake this sensitivity for "memory" of experience before regeneration.

All these difficulties are multiplied by the fact that in these classical conditioning procedures, the worm's responses are subjectively recorded, and the responses are not always clear enough to provide a basis for good observation. The use of instrumental learning procedures, such as the T maze with earthworms, would eliminate the problems of recording responses objectively and avoiding sensitization. Attempts in this direction have thus far not been eminently successful.

We still have much to learn about these simple invertebrates, and until we know more, experiments of the sort described here will be fraught with difficulties and beset by inconsistencies. Nevertheless, the evidence at hand, especially on earthworms, makes it clear that we can conclude with some confidence that associative learning can occur at the level of the worms.

Learning in molluscs has now been well documented. Not only are there striking experiments with the octopus (pp. 46–48), but now exciting new possibilities have opened up in the investigation of the sea hare, *Aplysia* (page 40), in which it is possible to correlate the behavioral changes in learning with specific neurophysiological changes in individual neurons involved.

Similar electrophysiological possibilities are evident in work with the arthropods also, and these organisms are capable of excellent learning. Bees can readily learn to fly to dishes placed on blue paper, for example, and avoid other dishes on gray paper if only the blue paper has been associated with sugar water. And they go consistently to the blue paper regardless of its relative position, even when all dishes are empty. Cockroaches and ants can learn simple mazes quite rapidly as experiments have shown. In the case of the ants, which are much better learners than cockroaches, the maze was interposed between the nest and the food, and learning was accomplished typically within 35 trials (Figure 5.26). Apparently, the learning in this case was not highly general, for when the ants had to run from the food to the nest, they had to learn the true path all over again.

It is obvious by now that there has been relatively little experimental investigation of learning in the invertebrates, especially of the lower forms. We are still beset by our lack of knowledge of these simpler organisms, so that on the one hand it is difficult to test them under optimal conditions and, on the other hand, it is difficult to avoid experimental artifacts or errors in interpretation. So far, positive evidence for learning is forthcoming only when we reach the level of the worms, which possess a bilaterally symmetrical, synaptic nervous system. Even more consistent and more complex learning is possible in the cephalopods and arthropods, with their large concentrated ganglionic masses, especially at the anterior pole of the central nervous system. Compared to the vertebrates, however, the higher invertebrates have relatively poorly developed learning capacity. They are quite stimulus bound, and their behavior in general is still dominated by the stereotyped patterns dictated by innate taxes, reflexes, and instincts.

We have already described many instances of vertebrate learning, so our treatment here may be brief and in rather general terms. Most of our knowledge is based on mammals, less on fish and birds, and the least on reptiles and amphibia, which have been very little studied. Fish have been successfully trained in a variety of discriminations involving simultaneous choice between two different objects. They have mastered mazes and detour problems much more rapidly and more consistently than invertebrates, with the possible exception of ants. Similar data have been obtained in a few studies on frogs and turtles. Modifiability and the ability to profit from experience, therefore, are clear characteristics of these animals. But by and large, their learning consists of trial-and-error learning of relatively simple stimulus-response relationships.

Birds show considerably more facility in learning than do the lower vertebrates, not only in the rapidity and consistency of learning, but also in the complexity of problems they are able to solve. In mammals, still further improvements are obvious as we compare simple mammals such as the rat and the subhuman primates. By this phylogenetic stage, new behavioral capacities emerge that increasingly free the animals from simple stimulus-response, trial-and-error learning. Many items from past learning experiences are retained and utilized in new learning tests. The higher mammals are not as stimulus-bound and as slavishly restricted by habits as are the lower vertebrates. In fact, they bring to the solution of complex problems, as we shall see later in this chapter, rudimentary capacities for reasoning and symbolic behavior that put much of their learning on a rather different basis from that of the simpler organisms. Man, of course, is enormously liberated from the need to approach each problem in a simple trial-and-error manner. His symbolic capacity in the form of language allows him to call on accumulated past experiences rapidly and extensively and to profit from symbolic communications from others in the learning he undertakes.

Obviously, the comparison of learning ability among different species is a difficult task. At the same time that learning ability is improving in phylogeny, improvements in sensory capacity and manipulative ability are giving higher animals additional advantages, and eventually, the capacities to solve problems by some form of rudimentary reasoning emerge as additional advantages of the higher animals.

If we were to attribute the improvement of learning ability to one thing in the phylogenetic series, it would be to the evolution of the central nervous system. It is an article of faith that learning represents some change in the central nervous system, and that memory is the preservation of that change.

Two simple examples of change induced in the nervous system will help make clear what the scientific problem is. In the first, a focus of irritation is produced on the cerebral cortex of a monkey by implanting aluminum cream on the cortical surface. This produces an epilepticlike discharge of nerve cells under the aluminum cream and, indeed, a mirror focus on the other side of the brain. At first the mirror focus discharges only when the original focus discharges and is completely dependent upon it, for if the two hemispheres are separated surgically, the mirror focus is abolished. After a few weeks, however, the mirror focus may discharge spontaneously and independently and at this point, the two hemispheres may be separated surgically and the mirror focus will continue to discharge. Evidentally some change occurred in the nerve cells of the mirror focus over a period of time that represents a kind of modification of function we can expect to see in learning.

The second example is similar. It is known that a lesion on one side of the cerebellum will produce a postural asymmetry so that the limbs on the one side are held flexed and those on the other are extended. Severance of the spinal cord right after such a cerebellar lesion destroys the asymmetry, for the unbalanced influence of the cerebellum is removed. However, if the unbalanced influence is allowed to continue for about an hour, then the asymmetry will persist after spinal cord severance. In other words, with time a lasting imbalance can be impressed on the spinal mechanism controlling the posture of the limbs.

In both these cases, the part of the nervous system undergoing change can be specified, and there is hope that the nature of the change can be studied. Great interest centers around the possibility that chemical changes had occurred in the nerve cells that were modified, particularly changes in ribonucleic acid (RNA). It is still too early to be certain, but the concept is an interesting one, based on the idea that experience may be encoded on a complex molecule like RNA in the same way that genetic information is encoded on deoxyribonucleic acid (DNA). According to this idea, learning causes a modification in the structure of RNA so that, in its function as a template for the synthesis of protein, it produces some specific modifications of protein structure in nerve cells.

A number of experiments on rats and mice have shown increases in RNA or shifts in base ratios of RNA following experience or training, but it is by no means clear that these effects are due to the encoding of memory rather than simply to increased neural activity. In another approach, using planaria, tail sections of trained worms were allowed to regenerate in water containing an enzyme that destroys RNA, and it was found that they failed to retain previous conditioning as did tail sections regenerated in plain water. Still another experiment reports that the drug 8-azaguanine, which inactivates RNA, slows the learning of rats. Unfortunately, in both of these experiments the effects are rather small, and neither experiment is sufficiently well controlled.

Even more dramatic claims have been made regarding the "transfer of memory" from one animal to another by "encoded RNA." In one, cannibalistic planaria ate either "trained" or "untrained" planaria. In another, an RNA extract was made of the brains of trained rats and injected into untrained rats. In both cases, the claim was made that the behavior of the recipient animals was modified specifically as though they had had the original training experience. In both cases, however, controls were lacking and there was no evidence in either case that the RNA was actually modified or that it actually reached the nervous system when it was injected. These are very serious shortcomings when such startling claims are being made!

Far clearer are the results of a study in which protein synthesis, in the brains of mice, was inhibited by the injection of a drug called puromycin. When the drug was injected into the posterior region of the brain after the learning of a simple position habit in a Y maze, there was complete loss of memory of 2 days' duration or less, but no effect on memory of 6 days' duration. To eliminate such longer-term memory, it was necessary to affect most of the remaining cerebral cortex with puromycin at the same time. These basic findings are strongly supported by similar experiments in goldfish, and further studies with mice have shown that puromycin reduces the electrical activity of the hippocampus to almost nil.

It turns out, however, that it is not just the inhibition of protein synthesis that is important in these studies, for acetoxycycloheximide inhibits protein synthesis profoundly without affecting memory in mice. The critical thing appears to be that puromycin inhibits protein synthesis by producing an abnormal peptide, which the heximide does not. The current concept is that the abnormal peptide remains in the brain for long periods of time, inhibiting the expression of memory. This interpretation is suggested by the fact that memory can be restored after puromycin injection into the brain by subsequently injecting saline into the same sites, and it is strengthened by the finding that labeled abnormal peptides can be found in the brain for many days after administration of radioactive puromycin. Much remains to be discovered before our knowledge of the chemistry of the memory trace attains firm ground, but a significant start has been made.

Whatever the details of a molecular concept of memory, it must operate in a neurophysiological and neuroanatomical setting as a consequence of behavioral experience. Thus, there must be some neurophysiological event in the nerve cells that produces the chemical change. And the chemical change in turn must have a way of altering the function of the nervous system to produce the memory. Most likely, the crucial changes will show up at the synapse, for it is here that the interactions among nerve cells can occur and it is here that modification in the function of the nervous system can take place. Certainly the chemistry of synapses should be the chemistry of learning and memory, and today it is possible to obtain adequate samples of synaptic chemicals by selective centrifugation of synaptosomes.

Some further insight into the nature of memory mechanisms has been gained by direct, experimental examination of *temporal characteristics of the memory process*. In one study, rats were trained to run from one compartment to another to avoid an electric shock. They were given one trial a day, and after each trial, they received an electroconvulsive shock (ECS) through the head. Different groups of rats received the convulsive shock at different times after the learning trial: after 20 sec, 1 min, 4 min, 15 min, 1 hr, 4 hr, and 14 hr. If the shock came within 1 hr, there was virtually no learning, but if it came after 4 hr, learning was essentially normal. Apparently, memory takes time to "set" in the brain, and this consolidation requires at least an hour, which suggests that memory is a two-part process, comprising an early phase when memory is vulnerable to electric shock and a later phase when it is not.

Other experiments report that with simpler, one-trial learning, consolidation may be achieved in 20 to 30 sec. Still other experiments report evidence of consolidation times of more than 1 hr, using drugs and other agents than ECS to interfere with the laying down of memory. Indeed, the concept now developing is that memory is a multistage process in which the time required for different stages may range from a few seconds to a few days. Different mechanisms (electrophysiological, chemical, anatomical) may be involved at each of these stages, and different parts of the brain may be involved as well.

Support for this temporal concept of memory comes from other, rather different experiments. In one, it was found that surgical isolation of the hippocampus in cats could block memory only if the operation were performed within 3 hr after learning. In another study, after the vertical lobe (an association area) of the brain of the octopus was removed, the animal could not remember a training trial for more than 15 min. A very similar defect is produced in human patients who have suffered damage to both temporal lobes. Patients with this affliction can learn something simple and retain it for about 15 min to an hour, but after that time they forget completely and may not even remember having learned. Yet their lifelong memories are left undisturbed.

Many other approaches have been made in the investigation of the neural basis of learning. Studies in which learning and memory are tested after parts of the nervous system have been destroyed surgically reveal that no one part of the nervous system, including the cerebral cortex, is crucial. Such brain lesions may impair the capacity to utilize sensory avenues, and to make skilled movements, to pay attention, or to be motivated to learn, but the basic capacities for learning and memory remain essentially intact. It is as though learning and memory occur over widespread areas of the brain, so that damage to any one part would be ineffective. As a matter of fact, efforts to record the electrical activity of the brain during learning support this idea, for changes in electrical activity may be found at many different loci in the brain as training proceeds.

PROBLEM SOLUTION, REASONING, AND INTELLIGENCE

It is merely a small step from questions about learning and memory to questions about problem solution, reasoning, and even intelligence. The basic neurological and evolutionary mechanisms at work are the same, even though the behavioral processes are much more complex. In fact, because of the behavioral complexity, it is more likely that we shall see striking changes in phylogeny and see more clearly the brain and behavior relationship in the study of reasoning and problem solution.

Reasoning is the ability to solve complex problems with something more than simple trial-and-error, habit, or stimulus-response modifications. In man, we recognize this capacity as the ability to develop concepts, to behave according to general principles, and to put together elements from past experience into a new organization, quite independently of the particular physical form a problem takes or the specific sensory or motor elements involved in the situation. What do we observe in animals? Earlier, we have described many cases of complex learning, in which an organism was required to handle many stimuli at once and to make a complex sequence of discriminative responses. Now we want to see performance in situations where specific sensory cues and specific habits are not critical for the solution of a problem.

A number of techniques have been devised to investigate such problem-solution capacities in animals, and these have met with varying degrees of success. One of the oldest, and in some ways the simplest, of these techniques is the *detour problem* (Figure 8.8). Here an animal is blocked from direct approach to food he can see and smell by a barrier or some other arrangement. To get the food, the animal must first move away from it and thus make a detour from the direct path. Several questions are pertinent. Can the animal solve the problem on its first exposure to it, that is, without having to learn by trial and error? Can an animal eventually learn to solve the problem? Or does it fail or simply have occasional "accidental" solutions as a result of excited running about? Of all the animals tested in detour problems, only monkeys and chimpanzees show any degree of success upon first exposure to the situation. Many other animals have learned to perform detours after failing on their first few trials. The octopus, for instance can slowly learn to detour under some conditions, but not under others. Fish and birds eventually learn the long way around a barrier. Laboratory-reared rats, dogs, and raccons are rapid learners, although they fail initially.

Making a detour on first exposure is similar to what we call *insight* into a problem in man. Even more direct evidence for insight-type solutions can be seen in other kinds of experiments with chimpanzees and, to some degree, with monkeys. These studies, first performed by Wolfgang Köhler during World War I, showed, for example, that a chimpanzee is capable of attaining a banana that is out of reach by stacking boxes beneath it and climbing up (Figure 8.9). Or a chimpanzee will fit two sticks together to pull in a piece of food that is out of reach of either stick alone. Unfortunately, it is difficult to be certain in these cases that the animal did not have some experience with a similar type

Bait

Figure 8.8   A detour problem for the raccoon in which the animal must first go away from the bait in order to reach it. [Redrawn from N. R. F. Maier and T. C. Schneirla, *Principles of Animal Psychology.* New York: McGraw-Hill Book Company, 1935.]

Figure 8.9 A chimpanzee solves the problem of getting a banana that is out of reach by stacking boxes on top of one another. [Redrawn from N. R. F. Maier and T. C. Schneirla, *Principles of Animal Psychology.* New York: McGraw-Hill Book Company, 1935.]

of problem before. Even so, the immediate application of past experience in a new situation is a noteworthy capacity.

To test for the application of past experiences to the solution of a new problem in a rat, the animal is taught two elements necessary for the solution of a problem and then is observed to see if it can combine them satisfactorily. For example, the rat is taught to climb down from a table to the floor, where it can explore the room (Figure 8.10). This is the first element. The second element is to train the animal to climb up another table and cross a runway to find food on a corner of the first table that is fenced off from the rest of it. When this is learned, the animal is placed on the first table on the opposite side of the fence from the food. Once it has mastered the two separate elements, the rat is able to climb down from the first table, run across the floor to the second table, climb it, and reach the food behind the fence. This, of course, is a detour problem in which the animal is trained in the elements of the solution beforehand and is required to put them together for the first time when the test is made.

Other approaches in testing the ability to solve problems in various members of the animal kingdom have required animals to learn to perform a task according to some general principle, quite aside from the specific stimuli available at the time the response is made. These animals are given a great deal of training. The major question is whether they can learn to master the problem at all. If they do learn, there is the question of how rapidly they can master it and at what level of complexity they can perform. One of the easiest of these problems is the *conditional reaction*. In this case, an animal might have to learn to choose the left of two gray doors if it is preceded by a single black door or the right door if it is preceded by a single white door; or to choose a door with the triangle instead of a door with the circle if the background has horizontal black and white stripes, and the one with the circle instead of the one with the triangle if the stripes are vertical. This is an "if–then reaction"; if the stripes are horizontal, then the door with the triangle is correct. Rats can gradually learn this type of response when required to make appropriate left-right choices in a runway divided alternately by single-cue doors (black or white) and double-choice doors (gray). Monkeys learn conditional reactions more readily, and are more versatile and stable in their solutions.

A more difficult type of principle-learning is what Harry F. Harlow has called the *learning set* (Figure 8.11). Here an animal is first required to discriminate between two common objects (say, a white jar lid and a red pincushion) that differ from each other in a number of physical properties. Displacement of the correct object, regardless of its position, yields food to the hungry subject. A monkey, for example, may require 50 to 100 trials to master such a problem. After this, other, different pairs of objects are presented for discrimination training. When a number of these have been learned, each succeeding new pair of objects is presented for only five or six trials. After 200 to 300 such six-trial presentations, the monkey can learn to make each new discrimination in one or two trials.

By chance, it can get the first trial correct only half the time. When it "guesses" correctly on the first trial, it will choose correctly in the remaining five trials and ignore the incorrect object. When it "guesses" incorrectly, it typically shifts to the other object and never goes back to the incorrect one. Thus, the first trial is an "information" trial upon which succeeding choices are based, as though in accordance with the principle: "If the first choice is correct, continue to choose that object and ignore the other; if it is incorrect, never go back to it." In essence, the animal has been trained to the point of "one-trial" learning, and this is similar to the insight case. Only primates have any degree of success in this type of problem. Rats fail miserably and must learn each new problem by trial and error, never carrying over the general principle from problem to problem.

Very similar but even more difficult is *oddity-principle learning*, in

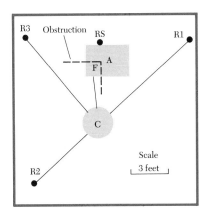

Figure 8.10   *The problem for the rat is to get from Table A to food at F when an obstruction blocks it way. First the rat is trained to go from A to the floor by way of a ladder down a ringstand, RS. Later it is trained to climb from the floor up ringstands R1, R2, or R3 and to reach Table C over the runways. From C it can get to F over a runway. The test is to see if the rat can combine two separate experiences and go directly from A to the floor to C to F on the first test.* [Redrawn from N. R. F. Maier and T. C. Schneirla, *Principles of Animal Psychology.* New York: McGraw-Hill Book Company, 1935.]

Figure 8.11   *The Wisconsin General Testing Apparatus for testing discrimination, learning set, and the solution of related problems in the rhesus monkey.* [From H. F. Harlow, in *Comparative Psychology*, 3rd ed., C. P. Stone, ed. Englewood Cliffs, N. J.: Prentice-Hall, Inc., 1951.]

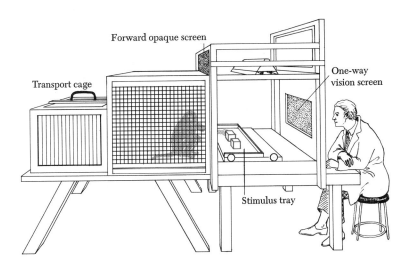

Figure 8.12 *A rhesus monkey performing an oddity-principle test.* [From H. F. Harlow, in *Comparative Psychology*, 3rd ed., C. P. Stone, ed. Englewood Cliffs, N. J.: Prentice-Hall, Inc., 1951.]

which three objects are presented, two the same and one different (Figure 8.12). The task is to pick the one different object, the odd one. Monkeys can master this task to the point where they are correct on the first trial of a new trio of objects on most occasions. Again, rats must learn each task anew and never reach the point where they succeed on the first trial (except through accident).

Still another type of problem is the *delayed reaction*, in which a hungry animal is presented with two identical cups. It is allowed to see food placed under one cup and then, after some delay, is released to make its choice by displacing one cup. With odor controlled, the animal has no discriminative cue at the time of choice between two identical cups other than its memory of which one was baited. To be used effectively, this memory must be carried over the period of delay, and it has been suggested that to do this the animal must be capable of some "symbolic process," akin to language, which it can use to represent the missing discriminative cue at the time of choice. Animals like the rat, cat, and dog solve this problem by orienting toward the baited cup during the delay and then "following their noses" to the correct cup. If they break orientation during the delay, or if the experimenter disorients them, they fail. Thus, they do not seem to show evidence of symbolic capacity. Raccoons, interestingly enough, are capable of short delays without orientation. Primates, however, are capable of much longer delays than rodents and carnivores and, what is more important, they do not maintain an orientation toward the correct stimulus during the delay period. They may even be removed from the test situation and returned after the delay and still perform successfully.

Another kind of symbolic process, akin to counting, is thought to be involved in the *double-alternation test*. Again, a hungry animal is presented wih two identical stimuli covering food wells, but this time the correct one is given simply by the order RRLL; that is, the food appears twice on the right and twice on the left, and it is this sequence that the animal must master. Rats fail this test. Cats will master a simple RRLL sequence with prolonged training, but they cannot extend the series; raccoons have been able to extend it as far as RRLLRR. Subhuman primates have extended the series to RRLLRRLL, but it is not until we get to man that indefinite extension of the series becomes possible.

Even more complex and more arbitrary sequences may be required of animals in the *triple-plate problem* (Figure 8.13). Here the animal simply has to press three pedals in some predetermined order, including the possibility that some pedals might have to be pressed more than once, not necessarily in succession. Pressing the correct sequence of pedals automatically opens a door to food. This test has been successfully used with guinea pigs, rats, kittens, and rhesus and cebus monkeys. Guinea pigs and rats are capable of pressing up to two or three pedals in order; kittens can handle as high a sequence as seven; monkeys, sequences of fifteen to twenty.

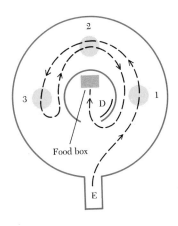

Figure 8.13 The triple-plate problem requires an animal to push one or all of three plates (1, 2, 3) in some sequence to open the food box. This test is useful in comparing the ability of different mammals to learn sequences because, first, the task of pushing plates is just about equally simple for all mammals and, second, the sequences can be arranged from the very simple to the very complex. The sequence here is difficult, requiring the animal to push plates 1, 2, 3, and 2, in that order, before going to the food box: D indicates door to food; E, entrance. [Redrawn from N. L. Munn, The Evolution and Growth of Human Behavior. Boston: Houghton Mifflin Company, 1955.]

Given these tests of problem solution involving insight, reasoning, concept formation, principle-learning, and symbolic processes, we are now faced with the question about what parts of the brain might be critically involved. Because performance on these tests improves so dramatically with the higher animals and, indeed, is sometimes possible only in the primates, the more recently evolved parts of the brain—the cerebral cortex and particularly its association areas—must be involved. Brain lesion studies in animals show the importance of the frontal association areas in delayed reaction and double alternation performance and of associational areas of the temporal lobe in learning sets and problems

LEARNING AND INTELLIGENCE

involving visual recognition. It turns out, however, that these association areas are not crucial for the tests used, for with sufficient training or with increased motivation or attention, the deficits produced by the lesions can largely be overcome.

It is not until we get to studies of brain injury in man that we can gain any real appreciation of brain function in complex intellectual processes. This is partly because animal experiments have not progressed far enough in investigating brain function in reasoning and partly because only man has reasoning processes (especially symbolic processes) that are developed far enough to provide a good basis for study.

The literature on brain injury in man is voluminous, but we shall only be able to cover some of the highlights, with emphasis on those that are pertinent to problems raised in the study of animals. Many studies indicate that intellectual deficit may follow lesions of the frontal lobe. This is clearly seen in cases of psychosurgery in which bilateral damage is done to the frontal lobes in order to control insanity or intractable pain and where the surgical damage tends to be somewhat back from the tips of the frontal lobes. Although the intellectual impairment may show itself in many ways, including inability to perform the double alternation test, it can generally be characterized as a loss of abstract as opposed to concrete ability. For example, the patient may be unable to "make believe" he is drinking from an empty glass or to go through the motions of writing his name with imaginary chalk on an imaginary blackboard. He is unable to carry out an idea in the abstract. If water is put in the glass, however, or if he is given chalk and a blackboard, he can follow the instructions normally, showing that he has the requisite abilities in the concrete situation.

When the brain injury is in the visual associational area, the patient may have a visual *agnosia*, which is inability to recognize common objects visually although he can see quite well and can recognize the objects through other senses. With parietal lobe injury, the agnosia may be tactual as well as visual, even to the point where the individual may not be able to recognize part of his body, such as his hand, as his own. In some cases, the defect is on the motor side (*apraxia*), and the patient is unable to carry out a meaningful movement such as hammering a nail, even though his motor reflexes are quite normal.

Perhaps the most striking of all the effects of brain injury in man are the *aphasias,* or language disorders. These take different forms, depending on the region of damage. If the damage is anterior in the cortex, the result will be a motor disorder such as the inability to name objects upon request; this result is seen after damage to Broca's area at the bottom of the motor cortex. Or it may take the form of an *agraphia*, an inability to respond in writing. More posterior lesions result in sensory aphasias—

either auditory, in which there is impaired comprehension of the spoken word, or visual, in which there is impaired comprehension of the written word (*alexia*). In all these cases, the individual's sensory and motor capacities may be quite normal, and his general comprehension may be good, but he simply blocks and gropes unsuccessfully, much as the normal person does when he cannot think of someone's name.

Much is yet to be learned about reasoning and symbolic capacity and their neural mechanisms in man, and, as we have pointed out, our knowledge of animals is even more deficient. But what we have learned so far shows that there is continuity from animal to man, even though man's specialization is so great that the step from animal to man is an enormously large one. Although only the mere rudiments of reasoning and symbolic processes are observed in animals, these processes are definitely there, and in animals such as the subhuman primates, they contribute to behavior in a way that clearly marks them off from the subprimate animals. So, although it is dangerous to overestimate the ability of animals by attributing human capacities to them, it is also dangerous to underestimate animals, especially the primates, or to try to account for their behavior solely in terms of simple stimulus-response associations.

Our conclusion, then, is that with the great development of the cortex in the subhuman primates and man, particularly of its association areas and related thalamic nuclei, new capacities for behavior emerge in evolution that fall under the general category of reasoning, abstract ability, and symbolic process. When these are developed, an organism is freed from slavish dependence on sensory stimuli, instinct, or habit and can make its adaptations to the environment with insight, reasoning, and the use of symbols, and ultimately of language.

SUMMARY

We have reviewed, in this chapter, the basic facts of animal learning and intelligence. We began with the concept that learning represents an *enduring modification of behavior* brought about by experience. Then we described various kinds of learning from the simple to the complex: habituation, classical conditioning, instrumental conditioning, and trial-and-error learning. The essential modification in behavior in all these cases is the development of some new response to a stimulus that never before elicited that response. As to the critical elements of the experience in learning, they appear to be very much the same in all these cases. Or put another way, these various instances of learning, including human verbal learning, all seem to obey the same fundamental laws of learning: contiguity, repetition, reinforcement, and, for the case of extinction or forgetting, interference.

When we took up the *phylogenetic development of learning,* we found strong evidence for learning at the level of the worms, in whom the bilaterally symmetrical, synaptic nervous system first appears. Habituation in minutes has been reported in micrometazoa, and there is now good evidence for learning in echinoderms. In the cephalopods and arthropods, which have relatively large, concentrated ganglionic masses in the anterior regions of the nervous system, learning ability is much greater than in the worms. Finally, with the development of the vertebrate brain, learning capacity develops even further, gradually reaching an asymptote among the simpler mammals (see Figure 7.1).

Despite this evidence that relates learning ability to the phylogenetic development of the central nervous system, it has not been easy to discover, with any degree of specificity, the neural basis of learning. The evidence from experimental brain lesions and, more particularly, from studies that record changes in the electrical activity of the brain during learning suggests that learning takes place in many places within the brain at once. The nature of the neural change in learning has proved elusive, however, even though recent experiments point to the possible importance of RNA and protein synthesis in the formation and maintenance of memory. Memory formation is probably a multistage process, different stages being mediated by physiological, biochemical, and anatomical changes and requiring anywhere from a few seconds to a matter of days. The long-term maintenance and storage of memory probably involves perhaps still other mechanisms, for it is possible to impair the permanent laying down of new memories by brain lesions without impairing either old, longstanding memories or the temporary acquisition of new ones.

We also have described the emergence of *reasoning and the symbolic process* in phylogeny. With the development of the associational areas of the cerebral cortex, the organism becomes free of strict stimulus control of its behavior. It is no longer a simple creature limited to instinct and habit, but is capable of insight, the formation of concepts and principles, and the use of symbolic processes in the solution of problems. Compared to man, animals possess only rudimentary powers of reasoning and symbolic behavior, and, to a large degree, such behavior does not appear until the development of the primates.

WE BEGAN THIS BOOK BY POINTING OUT THAT NO ANIMAL LIVES ALONE, and it is this fact that makes the study of animal behavior so intriguing and so important. While the individual organism may depend upon the adaptive mechanisms of instinct, learning, and reasoning to survive, these almost always operate in a social context, and moreover, survival of the species frequently depends upon some form of social behavior. Quite clearly, a social organization affords its own adaptive value for the individual and the species by facilitating reproduction and rearing of the young, by providing warning and additional defense against predators, and by contributing to the solution of the problems of food, water, temperature, and other challenges of the environment. At the same time, a social organization produces its own problems, contributing to emotional stress and disease. Nowhere in the animal kingdom are the benefits and the liabilities of a social organization so clearly illustrated as in the case of man and his societies. By looking at the evolution of social behavior and by analyzing the biological and behavioral mechanisms underlying it, we may be able to gain insights into the basic purposes and limits of society, its strengths and disorders.

Social behavior, by definition, implies the interaction of two or more individuals, or the influence of one individual on another. The mere existence of group behavior, of course, does not guarantee social behavior. For example, the orientation and movement of a group of paramecia toward a source of light is simply an aggregation of individuals responding to a common external stimulus. So are the circling of a group of moths around a lamp and the feeding of flies on a lump of sugar. Each individual would respond in the same way even if alone.

When we look back over the animal kingdom and examine what is involved in true social behavior, even in the simplest cases we find some elements in common that pose the major questions for our discussion. (1) To some degree, social interaction is involved in the *reproductive behavior* of most animals, the male-female and parental-filial interactions. (2) In many organisms, early *developmental history* is critical in setting the stage for later social roles, whether this be through physiological events that determine morphology (for example, bee class, sex) or psychological events that determine response tendency (for example, dominance, submission). (3) Most species have some means of *communication among individuals,* from a simple sign stimulus that arouses an instinctive pattern, to a vocal, symbolic process such as language. (4) *Territorial arrangements,* centering around breeding ground, nest site, and food source are common and provide a goal for many aspects of social behavior. (5) Finally, each species has a *characteristic social organization,* determining whether it will manifest permanent mateships, prolonged care of young, family, group or herd organization, socially induced disease, and cultural transmission from one generation to another; to a large degree, these characteristics depend upon the rate of development of the young and the neurological and endocrine endowment of the species.

REPRODUCTIVE BEHAVIOR

It is instructive to look at reproductive behavior, for at least in this one respect, almost all multicellular animals are social. The grayling butterfly shows no social behavior in its life cycle until it is time to mate. Then it begins an elaborate courtship. The male stops foraging and waits alertly for the female. Each time a female passes, the male flies in pursuit. If she alights and remains motionless, he jerks his wings and opens and closes them, exposing their spots. Then he does a deep bow, walks around her and brings his abdomen in contact with her copulatory organs. After this social behavior stops; the male never interacts with others again, and the female lays her eggs on food objects in the grass and leaves them.

An even more elaborate pattern in the stickleback fish was described (page 99), in which the male builds a nest, fights off intruding males, entices a female into the nest, induces her to lay eggs, fertilizes the eggs, cares for them by fanning them, and guards the young for a period of time after hatching. In birds like the herring gull, pairs remain together for longer periods, establish a territory, cooperate in building a nest, and go through patterns of courtship and copulation repeatedly. The male fights and threatens other males with highly specific postures, gestures, and calls. When the eggs are laid, the male and female take turns incubating them, one releasing the other with a highly specific approach to the nest. After the young are hatched, the cooperation extends to feeding and guarding them. When predators enter the colony, the adults sound the alarm call and fly up, and the chicks run for cover. The birds then swoop down upon the intruders and "bomb" them with regurgitated food and feces.

As we saw earlier, these interactions among individuals are largely innate patterns, which are heavily dependent on hormones and which are aroused by highly specific sign stimuli provided by another individual. Sometimes the sign stimulus is a morphological change induced by hormones, such as the brilliant coloration of the male stickleback. Sometimes it is a behavioral response, such as the bow of the grayling. Typically, the sign stimulus and the response to it are parts of a complex chain, such as the alarm reaction to an intruder that brings forth a warning call, which in turn arouses flight in individuals who have not been stimulated by the intruder. To some degree the effectiveness of a sign stimulus may depend on learning, as in imprinting. In this case, the innate following response is elicited specifically by the individual or object present at the time of hatching.

The intricacy and delicate balance of these complex mechanisms guiding mating and parental behavior in the reproductive cycle are illustrated beautifully in the study of the ringdove. Here we see that both behavioral and physiological factors work to elicit and guide the interactions of the male and female early in the reproductive cycle and of the parents and offspring later in the cycle. Each episode of courtship behavior of the male dove excites marked ovarian growth in the female dove and thus causes an increase in sex hormone levels; this in turn leads to a behavioral sequence of nest building, mating, egg laying, and egg incubating. Even a male separated from the female by a glass partition has this effect, providing he displays courtship behavior. If he is castrated, he does not bow and coo and does not affect the female; but if the castrate is injected with male sex hormones, he will exhibit the courtship behavior and thus will be able to stimulate ovarian growth in the female.

In a similar fashion, the process of incubation of eggs leads to a great increase in the size of the crop in both the male and the female, and crop milk is produced for the young when they hatch. Within a period of about 2 weeks, when the young have been weaned, the adult male begins to bow and coo again and another round of reproductive behavior is started.

Thus we have a behavioral mechanism for synchronization of the separate hormonal cycles in the two sexes of ringdove, presumably through an optic-hypothalamic, pituitary-gonadal pathway dependent on the female's sight of the male. At the same time, the gonadal hormones released into the internal environment selectively arouse hypothalamic mechanisms that facilitate the elicitation of mating or parental behavior. Yet it is clear that still other mechanisms operate in the reproductive cycle of the ringdove since the cycle is quicker and surer with experienced birds than with naïve birds. In addition, the whole process is facilitated by auditory stimulation provided by colony members, which is controlled by the size of the colony in the wild (the larger the colony, the greater the fertility of mating pairs) or by the volume control of a loudspeaker playing tapes of colony sounds into a laboratory cage.

Similar studies in mammals and even subhuman primates demonstrate similar biological and behavioral bases for sexual, parental, and filial behavior. For example, the maternal behavior of the rat is dependent on internal hormonal secretions, on the stimulation provided by the young, and possibly on past experience. The virgin female will not typically retrieve young or tuck them under her for nursing. However, if each day a virgin female is provided with a new set of baby rats several days old, within 6 days she will exhibit strong maternal behavior. Of course, in nature, a virgin female rat would never encounter 6 days of experience with infant rats that stayed the same age, but the experiment lets us see, in an exaggerated way, the effect that the young can have in producing maternal behavior. This becomes important when subsequent experiments reveal that infusion of blood from a lactating female rat into a virgin female causes the virgin to show the maternal response to the infant rats in 1 or 2 days, rather than in 6.

Obviously, the hormonal stimulus facilitates the neurological mechanism through which the social stimulus provided by the young works. Indeed, direct arousal of the central neurological mechanism in the brain has been demonstrated by injecting hormones into the brains of rats and ringdoves and by electrical stimulation of regions of the brain concerned with maternal behavior. In these cases, prompt and effective maternal responses to the young can be evoked when they would not otherwise appear.

Although an individual or a species may have an innate capacity for a particular kind of social interaction, a great deal depends upon early experience. Indeed, some aspects of social behavior so clearly depend on early experience that either they will not appear or they will be severely distorted if the experience is lacking. We have already seen that the chaffinch must hear the song of its species before it can sing its complete pattern and thus use it in normal, adult communication (page 97). Moreover, it must hear the song before its first breeding season, so that there is a critical period for the development of this facet of social behavior.

The same kind of critical period in social development shows up early in the life of many organisms. Inprinting in birds is a classical example, in which there is less than a day in which the hatchling can develop a strong following response to a mother or mother substitute

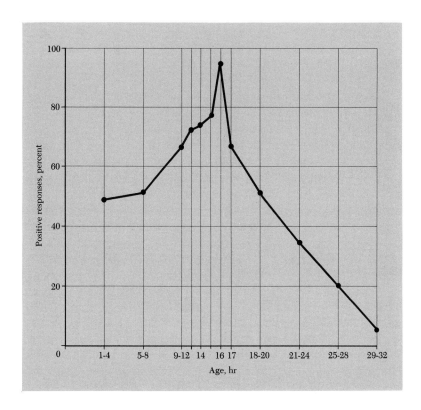

Figure 9.1 *Critical period for imprinting is less than a day after hatching.* [From E. H. Hess, in *Psychobiology: The Biological Bases of Behavior,* J. L. McGaugh, N. M. Weinberger, and R. E. Whalen, eds. San Francisco: W. H. Freeman and Co., 1967, p. 109.]

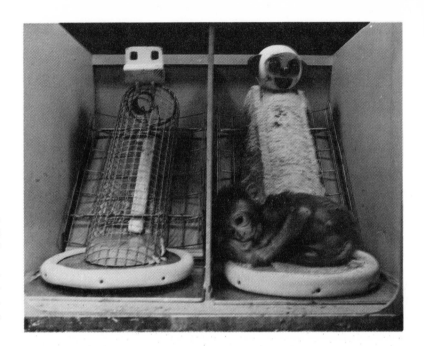

Figure 9.2   Infant rhesus monkey showing preference for cloth and foam-rubber mother substitute over hard, wire substitute. [From H. F. Harlow, in *Psychobiology: The Biological Bases of Behavior,* J. L. Mc-Gaugh, N. M. Weinberger, R. E. Whalen, eds. San Francisco: W. H. Freeman and Co., 1967, p. 101.]

(Figure 9.1). In sheep, the lamb will be accepted by the mother only within the first 4 hr after birth.

Even more extensive studies have been done on monkeys that were reared with various kinds of artificial, nursing mothers (Figure 9.2). Compared to the young monkey reared with its own mother, young reared on artificial, inanimate mother substitutes are timid in response to strangers and in strange situations, they do not play adequately with their peers, and they do not mate successfully when they become mature. If the artificial mother is made of hard wire material, these deficiencies in social behavior are severe and disabling; if the artificial mother is made of soft terrycloth and foam rubber, the symptoms are far less marked. What is more, the young develop even better if the artificial mother is made to rock or swing rather than remaining fixed and still.

Obviously, tactile stimulation and motion are both important components of the infant-mother relations that contribute to the development of normal social behavior in the rhesus monkey. Clinical studies suggest that the human infant may be even more dependent upon certain critical aspects of the interaction with its mother since infants reared under excellent nutritional and health conditions in orphanages may suffer from lack of affection. When this happens, they fail to thrive and often succumb to disease; if they do survive, they are often dull and socially withdrawn.

Other facets of social behavior also depend upon early life experiences. In one study of a group of rats that were kept in a large outdoor pen, it was found that offspring of submissive rats became submissive as a result of experience administered to them by dominant rats. In this social situation, the dominant rats lived near the central food and water supply and the submissive rats lived out in the perimeter, going to the food bin only during the day when the dominant rats slept. At weaning time, the naïve offspring of these submissive rats wandered directly into the territory of the dominant rats, and the dominant females would seize them, hold them down, and "drub" them with their paws without inflicting actual injury. After a few such treatments, the offspring of submissive rats became submissive too. Comparable examples in man are too easy to think of to need listing.

Common observation shows that many animals maintain territories which they will defend, but not until detailed studies of English song birds were made early in this century was it appreciated how complex and how intense territoriality might be. The song sparrow, for example, takes up a "display station" in a territory that may be as large as an acre,

Figure 9.3 Drawing made from a photograph, showing the small nesting territory of gannets. [From W. Etkin, Social Behavior and Organization Among Vertebrates: Types of Social Organization in Birds and Mammals. Chicago: University of Chicago Press, 1964.]

a

b

Figure 9.4 (a) Herring gull declaring nest site by displaying "choking" posture. (b) Gull defending its territory with "oblique" and "long call" displays. [From K. Z. Lorenz, in *Psychobiology: The Biological Bases of Behavior,* J. L. McGaugh, N. M. Weinberger, and R. E. Whalen, eds. San Francisco: W. H. Freeman and Co., 1967, p. 34.]

and there he sings his song with great persistence. If another male sparrow shows up, he is driven off aggressively. As spring progresses, the singing and competing for territory become more intense, and indeed, the territory may be divided in half more than once before the pairing with a female takes place.

Similar defense of territory may be seen in stickleback fish and in gulls and gannets, especially during the nesting period. These birds will establish nests very close to each other on a crowded strand or ledge and each bird will identify its own nest without error and respect the common boundary of its territory and its neighbors' (Figure 9.3). The gull asserts its territory with a clear display (Figure 9.4a). The intruder that crosses the territorial line is greeted with an aggressive display (Figure 9.4b), both postural and vocal, that very quickly induces retreat, but the animal that retreats is quite capable of the same effective, aggressive display when defending his own territory against an intruder. It is a common observation in many birds and mammals that a given individual will be aggressive or submissive, depending upon whether the social encounter occurs in its own territory or another individual's.

In many instances, the territory is not so much a defended area as it is a "home range." For example, repeated trapping of rats in a city show that 89 percent of the animals can be trapped again within 60 ft of the point where they were first trapped. The same has been observed in field mice and, over somewhat larger territories, in woodchucks. Antelopes and other ungulates may range over a territory of several square miles, and in this case it is the territory of a group or herd, at least until mating season, when the dominant males fight to set up an individual territory.

Troops of monkeys and chimpanzees, and families of gorillas as well, move through home-range territories that are clearly their own although not always closely defended. A baboon troop of 40 to 80 individuals may range over an area up to 6 mi² (square miles) and this territory may overlap the ranges of one or more neighboring troops, especially around a water hole. With as many as 400 baboons at a single water hole, it has been reported that there need not be any mixing of troops nor any fighting, although small troops will yield to larger ones.

Territoriality obviously serves many functions. It certainly facilitates mating and care of the young, but it also provides for habitual solution of the problems of feeding and maintaining shelter. Observation also shows that territoriality reduces aggression or actual fighting to occasional displays necessary to defend a well-established territory and drive out intruders. Finally, because territories are required for mating and the rearing of the young, it is also clear that territoriality is one of the factors holding population in check.

All of the interactions described in the examples of social behavior offered here involve communication of various sorts from one animal to another. Sometimes communication is a display, such as the "bowing and cooing" of the male ringdove. Sometime it is the complete reduction or elimination of aggressive display signals such as the "surrender response" seen in dogs, wolves, and rats, for example. These animals will roll over on their backs and present their bellies and necks when they are defeated in an aggressive encounter (Figure 9.5). The attacker no longer is stimulated by signals which evoke attack and he therefore stops his attack.

The sign stimuli of a display that controls the behavior of others are so important that they are frequently presented in formal, *ritualized* responses, which unfold slowly and with great emphasis so that no detail need be missed. The "zig-zag dance" of the stickleback is a good example and so is the "strutting" of the male sage grouse, who functions as the master cock and does 80 percent of the mating while other males stand guard in a circle, undoubtedly also controlled by the master's aggressive strutting ritual (Figure 9.6).

Communication need not always be visual or auditory, nor must it always be confined to the present. It may be chemical and because of this it may, of course, span time. Thus a male dog "marks" his territory for some time into the future and the intruder encounters a scent that has some history. The gerbil and the marmoset also do this, not only with urine, but also by rubbing scent glands located around the genitals or on

Figure 9.5 **Defeated rat rolls over on his back, causing victor to stop his attack (a) and sniff the motionless loser (b).** [From S. A. Barnett, *The Rat: A Study in Behaviour.* Chicago: Aldine Publishing Company; London: Methuen & Co. Ltd., 1963.]

Figure 9.6 (a) "Strutting" display of male sage grouse: the master cock. (b) Males "standing guard" on periphery to let waiting females into the area for master cock to mate. [After J. P. Scott, Animal Behavior. Chicago: University of Chicago Press, 1958.]

the chest. As we saw in Chapter 5 (page 67), specific chemicals involved in social communication are called *pheromones*, by analogy with hormones; these are small quantities of specific chemical materials acting upon another organism at a distance from their point of release and in very specific ways.

The term pheromone originally referred to the chemical influences social insects had on each other, from the influence of royal jelly in producing a queen to the recognition of strangers by odor or the following of "smell trails" by ants. It now has been broadened to mean any form of specific chemical influence produced by one organism and acting on another. Thus it includes such diverse examples as the marking of territories by urine and scent gland secretions, odors that function from a distance as sexual attractants in insects, odors from a strange male that can produce "pregnancy block" in female mice. In this last case, the introduction of a strange male after a female mouse has been impregnated can lead to a return to the estrous condition (spontaneous abortion), even 24 hr after impregnation. If the olfactory bulbs of the female are removed, however, the pregnancy block will not occur.

One of the most fascinating forms of communication is seen in the so-called language of the bees, described earlier (pp. 65–69). In addition to communicating through the form of its dances, the bee also emits a train of sound in each straight run of the waggle dance and the number of pulses in the train correlate highly with the distance to the food. The sound spectrography used in these studies also revealed at least 10 other sounds, some of which may also be important in communication. For example, when the hive is disturbed, the "hum" of the hive is interrupted by short bursts of sounds, produced when bees guarding the hive rock forward on their legs. These bursts may be repeated every few seconds for as long as 10 min. There then follows a "piping" of the workers, which is produced by faint beeps every half second as the agitation of the hive subsides. It is also thought that the queen bee communicates by sound, "tooting" upon first emerging from her cell and "quacking" when she is mature and trying to leave the hive. Through sound spectrography, each of these sounds can be analyzed as to frequency, intensity, and pattern, and thus in an objective way can be related to behavior.

Sound communication is of a special interest, of course, because it forms the fundamental basis of human language. Language as we know it requires (1) articulate production of sound and the capacity for auditory discrimination; (2) a basic brain capacity for symbolic processes in both the production and the appreciation of language; and (3) a social organization in which language is used for communication according to a set of rules that are passed on from one generation to another. The excellent articulation of the parrot or the mynah bird is not language, for their vocalizations are not symbolic and are not used as primary means of social communication. On the other hand, the vocalization of monkeys and other subhuman primates, while crude as sound productions, do communicate symbolically in the social setting. For example, it is possible to recognize from six to twelve different symbolic sounds in the social interactions of monkeys, conveying different kinds of warning or representing food cries.

Efforts to teach "words" to subhuman primates have been largely disappointing. It has been claimed that the chimpanzee can be trained to use the words *mama, papa,* and *cup* symbolically and correctly, but the training process is laborious and success is limited and almost nonexistent. On the other hand, chimpanzees have been trained to use gestures symbolically in their interactions with man and this seems to be much easier and much more productive. It is possible that the chimpanzee and other subhuman primates simply do not have the brain capacity to produce sounds on command and to relate them to the appropriate symbolic meaning. In this sense, they are like the human patient who has suffered damage to the speech areas of his brain (see page 130) so that he is aphasic and cannot produce language.

The gift of language alone sets man apart from the rest of the animals and gives him enormous advantages in his adaptation to his environment and in his social organization. Language is an unbounding symbolic tool of his intelligence; in man's social behavior, it is a mechanism for communication and for cultural transmission from one generation to the next.

Figure 9.7   A group of 40 to 50 submissive mice that remain huddled in a corner of the cage area most of the time. [From R. L. Snyder, in *Progress in Physiological Psychology,* E. Stellar and J. M. Sprague, eds. New York: Academic Press, Inc., 1968, Vol. II.]

SOCIAL ORGANIZATION

On the basis of innate patterns, remarkable social organizations have developed, perhaps reaching their peak in insects. Ants have wondrously complex social organizations, which contain specialized classes of individuals and in which there is cultivation of fungus "gardens," "milking" of aphids, and exclusion of outsiders to the colony. In the simpler vertebrates (for example, fish and birds), some social organization may be seen in migration, schooling, flocking, and herding. Some birds display a pecking order in which the dominant bird pecks the more submissive bird. Such pecking order gives a highly particular and elaborate social organization to a group of birds, determines which individuals will do the mating, and also serves the function of keeping aggression and physical injury to a minimum, once the order is established.

Similar dominance and submissive relationships determine not only the kind of social organization possible, but also some of the consequences of that organization for the health of individuals and the size of animal populations. In a study of mice kept in a large laboratory cage area (42 ft²) with a completely adequate supply of food and water, it was found that as the population grew from the original 2 male-female pairs, the older animals, the first arrivals, became the dominant individuals and were sleek in their coats, well fed, and responsible for most of the breeding. Two kinds of submissive animals were found in this study: "huddlers," which were mice that congregated in large numbers (up to 100) in the various corners of the cage area, lying one on top of the other (Figure 9.7), and "recluses," which were individuals that had found small perches on which they could sit and be isolated from the rest of the social environment (Figure 9.8). Both kinds of submissive animals fed only during the daytime when the dominant animals slept; they were thin, showed loss of hair, and often had skin lesions. These symptoms were the result of the great pressure put upon the submissive animals by the dominant ones in their various social contacts, presumably as a result of neuroendocrine changes in the submissive animals, especially adrenal hyperplasia and gonadal atrophy. Not only did these endocrine disorders lead to a high rate of mortality among the submissive animals, but they also quite obviously led to reduced fertility so that fewer and fewer young were produced in the population. Consequently, as Figure 9.9 shows, the population of mice declined, presumably until the social pressure was reduced so that the mortality rate was decreased and the fertility rate increased, resulting in a subsequent rise in population. Many animal populations in nature show similar periodic rises and crashes in population, extending over a 2- to 3-year period, and these appear to be more a matter of social pressure within a limited territory than anything else.

Striking details of the pathologic changes that can be caused by social pressure are seen in the laboratory study of chickens housed in different numbers and in different sex ratios. Heart disease, for example, increased as the group size increased beyond two. But when there were four or five males to each female, or vice versa, the increase in coronary-artery pathology and in mortality was very marked, although the males proved more susceptible than the females. Implications for man and his

Figure 9.8 *A recluse that remains isolated on a protected perch most of the time.* [From R. L. Snyder, in *Progress in Physiological Psychology*, E. Stellar and J. M. Sprague, eds. New York: Academic Press, Inc., 1968, Vol. II.]

Figure 9.9 *Rise and fall of a population of mice in a confined cage area over a 3-yr period. Note the relation of number of births to total population.* [From R. L. Snyder, in *Progress in Physiological Psychology*, E. Stellar and J. M. Sprague, eds. New York: Academic Press, Inc., 1968, Vol. II.]

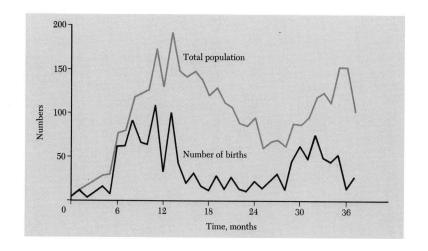

societies are clear. Social organization and social contacts that put emotional pressure and stress on individuals can result in disease and premature death. Much evidence today would indicate the role of similar social factors in many of the so-called degenerative diseases of civilization.

Social organization is one of the major modes of adaptation to the environment. To some extent, it appears at all levels of the evolutionary scale, but evolutionary development is not continuous nor along one line. For example, while the grayling butterfly interacts with another member of its species only briefly during mating, the social organization of bees and ants can be enormously complex; yet many vertebrates have minimal social relations and it is not until the subhuman primates and man are reached that there is much in the way of complex social organization.

Much of social behavior in the phylogenetic scale is dependent upon innate responses to specific patterns of social display, the *social sign stimuli*. Both the exhibition of the display and the response to it may depend, in turn, upon the hormonal state of the organism. Alternatively, they may depend on past experience, as in the case of imprinting.

Much of social behavior centers around *territory*, whether it be the defended territory of an individual or the "territorial range" of a group. Territoriality limits population size, minimizes injurious aggression, and fosters social organization based on dominance-submission relationships.

In early life development, there are critical periods for the occurrence of morphological or behavioral changes that are crucial to the process of socialization typical of the species, whether it be filial following in birds or affectional responses in monkeys.

Social communication among individuals ranges from simple sign stimuli and gestures to language. Symbolic capacities probably do not emerge until the level of the primates, and it is a huge step from the subhuman primate to man.

While social organization produces its own benefits in the adaptation of the individual and the species to the environment, it also clearly puts pressure on the individual. If this pressure is not adequately regulated, a variety of maladaptive results can occur, including abnormal behavior, disease, and death.

ADRIAN, E. D. *The Physical Background of Perception.* London and New York: Oxford University Press (Clarendon), 1947. A brief account, by one of the pioneers of modern electrophysiology, of the electrical activity of nerve cells and the ways in which this activity is coded to provide the basis of sensation and perception.

ETKIN, W., ed. *Social Behavior and Organization Among Vertebrates.* Chicago: University of Chicago Press, 1963. Original articles by the editor and other scientists in the field (F. A. Beach, D. E. Davis, D. S. Lehrman, J. P. Scott, N. Tinbergen, and others).

VON FRISCH, K. *The Dance Language and Orientation of Bees.* Cambridge, Mass.: Harvard University Press, 1967. A comprehensive monograph, fascinating and up-to-date, representing more than 50 years of investigation by von Frisch and his students.

HINDE, R. A. *Animal Behaviour.* New York: McGraw-Hill Book Company, 1966. An excellent modern text that brings together concepts from ethology and comparative psychology.

LORENZ, K. Z. *King Solomon's Ring: New Light on Animal Ways.* New York: Thomas Y. Crowell Co., 1952. A popular book by the founder of modern ethology, which describes the way in which careful field and laboratory observations permit the uncovering and understanding of many remarkable examples of animal behavior.

MCGAUGH, J. L., N. M. WEINBERGER, AND R. E. WHALEN, eds. *Psychobiology.* San Francisco: W. H. Freeman and Co., Publishers, 1967. Highly readable and stimulating articles from *Scientific American,* written by leading scientists.

MCGILL, T. E., ed. *Readings in Animal Behavior.* New York: Holt, Rinehart & Winston, 1965. Fifty-five selections by specialists representing a wide variety of disciplines. The articles originally appeared in professional journals.

MAIER, N. R. F., AND T. C. SCHNEIRLA *Principles of Animal Psychology.* New York: McGraw-Hill Book Company, 1935. A classical textbook of comparative psychology, systematically covering the animal kingdom from protozoa to primates.

MANNING, A. *An Introduction to Animal Behavior.* Reading, Mass.: Addison-Wesley, 1967. A readable introduction to animal behavior written primarily from an ethological point of view.

OCHS, S. *Elements of Neurophysiology.* New York: John Wiley & Sons, Inc., 1965. A simple yet comprehensive introduction, which will serve as a useful background for understanding neural aspects of behavior.

PARKER, G. H. *The Elementary Nervous System.* Philadelphia: J. B. Lippincott Co., 1919. A classical and highly readable account of the evolution of the structure and function of the nervous systems of the lower animals.

ROMER, A. S. *The Vertebrate Body.* Philadelphia: W. B. Saunders Co., 1962. One of the best treatments of comparative antatomy, with an excellent chapter on the vertebrate nervous system.

SCOTT, J. P. *Aggression.* Chicago: University of Chicago Press, 1958. An examination of the psychological, physiological, social, and ecological bases of aggression.

THORPE, W. H. *Learning and Instinct in Animals.* Cambridge, Mass.: Harvard University Press, 1958. An attempt to synthesize the approaches to the study of animal behavior employed by ethologists and psychologists; special attention is devoted to insects and birds.

TINBERGEN, N. *The Study of Instinct.* London and New York: Oxford University Press (Clarendon), 1951. A classical statement of the original position of European ethologists; provides valuable insight into the ethological method of approach.

———— *Social Behaviour in Animals.* London: Methuen & Co. Ltd., 1953. A classical account of patterns of social behavior in the animal kingdom.

WATERS, R. H., D. A. RETHLINGSHAFER, AND W. E. CALDWELL *Principles of Comparative Psychology.* New York: McGraw-Hill Book Company, 1960. A modern textbook written from the point of view of American psychology, with strong emphasis on learning.

WELLS, M. J. *Brain and Behavior in Cephalopods.* Stanford, Calif.: Stanford University Press, 1962. A complete account of behavior studies, with special emphasis on learning and its neurological correlates.